UNIVERSITY OF
WOLVERI

D0809744

WITHDRAWN

PROBLEMS IN PROBABILITY

2nd Edition

SERIES ON CONCRETE AND APPLICABLE MATHEMATICS

ISSN: 1793-1142

Series Editor: Professor George A. Anastassiou
Department of Mathematical Sciences
The University of Memphis
Memphis, TN 38152, USA

---

*Published**

Vol. 4 Stochastic Models with Applications to Genetics, Cancers, Aids and Other Biomedical Systems
by Tan Wai-Yuan

Vol. 5 Defects of Properties in Mathematics: Quantitative Characterizations
by Adrian I. Ban & Sorin G. Gal

Vol. 6 Topics on Stability and Periodicity in Abstract Differential Equations
by James H. Liu, Gaston M. N'Guérékata & Nguyen Van Minh

Vol. 7 Probabilistic Inequalities
by George A. Anastassiou

Vol. 8 Approximation by Complex Bernstein and Convolution Type Operators
by Sorin G. Gal

Vol. 9 Distribution Theory and Applications
by Abdellah El Kinani & Mohamed Oudadess

Vol. 10 Theory and Examples of Ordinary Differential Equations
by Chin-Yuan Lin

Vol. 11 Advanced Inequalities
by George A. Anastassiou

Vol. 12 Markov Processes, Feller Semigroups and Evolution Equations
by Jan A van Casteren

Vol. 13 Problems in Probability, Second Edition
by T M Mills

*To view the complete list of the published volumes in the series, please visit:
http://www.worldscientific/series/scaam

Series on Concrete and Applicable Mathematics – Vol. 13

PROBLEMS IN PROBABILITY

2nd Edition

T M Mills

La Trobe University, Australia &
Loddon Mallee Integrated Cancer Service, Australia

NEW JERSEY • LONDON • SINGAPORE • BEIJING • SHANGHAI • HONG KONG • TAIPEI • CHENNAI

Published by

World Scientific Publishing Co. Pte. Ltd.
5 Toh Tuck Link, Singapore 596224
USA office: 27 Warren Street, Suite 401-402, Hackensack, NJ 07601
UK office: 57 Shelton Street, Covent Garden, London WC2H 9HE

British Library Cataloguing-in-Publication Data
A catalogue record for this book is available from the British Library.

Cover design is based on the drawing by Linda Perry.

Series on Concrete and Applicable Mathematics — Vol. 13
PROBLEMS IN PROBABILITY
Second Edition

ISBN 978-981-4551-45-8

Printed in Singapore by World Scientific Printers.

To Frances, may the Lord bless thee and keep thee.

Preface

Probability theory is an important part of contemporary mathematics. The subject plays a key role in the insurance industry, modelling financial markets, and statistics generally — including all those fields of endeavour to which statistics is applied (e.g. health, physical sciences, engineering, economics, social sciences). The twentieth century was an important time for the subject because we witnessed the development of a solid mathematical basis for the study of probability, especially from the Russsian school of probability under the leadership of A.N. Kolmogorov. We have also seen many applications of probability from diverse areas such as stochastic calculus in the finance industry, genetics, fundamental ideas in computer science, queueing for surgery, and internet gambling.

Almost any aspect of mathematical modelling offers opportunities to apply probability theory. In the 21st century, the subject offers plenty of scope for theoretical developments, modern applications and computational problems. Probability is now a standard part of high school mathematics, and teachers ought to be well trained in the subject. Every student majoring in mathematics at university ought to take a course on probability or mathematical statistics.

Problem solving is important in mathematics. There are two distinctive features of the problems in this book. First, the problems tend to be quite long, verging on small projects. Many are intended to give the student the flavour of mathematical research even though the problems are not research problems *per se*. Long problems help to dispel the feeling that all mathematics problems can be solved in half a page. Second, some problems are well known theorems (e.g. the central limit theorem). Many texts give the student the opportunity to study the proof and to apply such theorems. The problems in this book give the student an opportunity to experience proving the theorem.

These notes and problems are designed to:

- provide a basis for a series of lectures suitable for advanced undergraduate students on the subject of probability,
- show, through problems, the excitement of probability,
- assist students to develop interest and skills in solving problems,
- introduce students to famous works and workers in probability,
- convey classical and contemporary aspects of probability,
- allow students to experience the style of thinking involved in mathematical research,
- improve students' library research skills.

Part I consists of notes and problems; Part II consists of complete solutions and related comments. The statement of a problem often contains the answer, a hint, or a source of further information. The solutions in Part II often contain discussion and suggestions for further reading. The book is not designed to be self-contained because it is my intention that students will be driven to some of the references listed in the bibliography and thereby encounter some of the classic works in probability.

A feature of this edition is the set of drawings of Linda Perry, an Australian artist, who received her training at East Sydney Technical College and La Trobe University. Linda describes these works as follows.

"Drawing goes beyond being a tool of description or recording; it also has intrinsic value as a spontaneous form of expression. Every mark or line contains emotional content capable of communicating feeling. These drawings explore contrasts; contrasts of lines and other marks, of spaces, of tones and of textures. In some of the drawings gentle lines appear, wander randomly across the space then disappear, while in others, accumulated, aggressive lines push and shove at other spaces. The images are the end result of printing marks that have been etched into metal plate. It is a process that can enhance the emotional content of marks and lines. It is an exciting and appealing extension to the drawing process itself."

Bendigo
June 2013

Contents

PART 1

Problems

Chapter 1

Sets, measure and probability

1.1 Notes

Logic

If p, q denote propostions then

$\neg p$ denotes the proposition "not p";
$p \wedge q$ denotes the proposition "p and q";
$p \vee q$ denotes the proposition "p or q";
$p \Rightarrow q$ denotes the proposition "p implies q";
$p \Leftrightarrow q$ denotes the proposition "p implies q and q implies p";
$(\forall x)(p)$ denotes the proposition "for all x, p is true";
$(\exists x)(p)$ denotes the proposition "there exists an x such that p is true".

Sets

$x \in A$: x belongs to the set A; x is an element of the set A
$x \notin A$: x is not an element of the set A
$A \subset B$: A is a subset of B; $x \in A \Rightarrow x \in B$
$A = B$: $A \subset B$ and $B \subset A$; $x \in A \Leftrightarrow x \in B$
$A \cup B = \{x : x \in A \text{ or } x \in B\}$
$A \cap B = \{x : x \in A \text{ and } x \in B\}$
$A' = \{x : x \notin A\}$
$A \setminus B = A \cap B'$
$A \triangle B = (A \cap B') \cup (B \cap A')$

Commutative laws:

$$(A \cap B) = (B \cap A)$$
$$(A \cup B) = (B \cup A)$$

Associative laws :

$$(A \cap B) \cap C = A \cap (B \cap C) = A \cap B \cap C$$
$$(A \cup B) \cup C = A \cup (B \cup C) = A \cup B \cup C$$

Distributive laws :

$$(A \cap B) \cup C = (A \cup C) \cap (B \cup C)$$
$$(A \cup B) \cap C = (A \cap C) \cup (B \cap C)$$

Complementation laws:

$$(A \cap A') = \emptyset$$
$$(A \cup A') = \Omega \text{ (where } \Omega \text{ denotes some universal set)}$$

De Morgan's laws :

$$(A \cap B)' = (A' \cup B')$$
$$(A \cup B)' = (A' \cap B')$$

Measure and probability

We begin with the definition of a σ-algebra of sets.

Definition 1.1 Let Ω be a set and $\mathcal{A}$ be a set of subsets of Ω. Then $\mathcal{A}$ is a σ-algebra if

- $\Omega \in \mathcal{A}$,
- $A \in \mathcal{A} \Rightarrow A' \in \mathcal{A}$, and
- $\{A_1, A_2, A_3, \ldots\} \subset \mathcal{A} \Rightarrow \cup_{i=1}^{\infty} A_i \in \mathcal{A}$.

We now define a probability measure.

Definition 1.2 Let Ω be a set and $\mathcal{A}$ be a σ-algebra of subsets of Ω. The function $P : \mathcal{A} \to [0, 1]$ is a probability measure if

- $P(\Omega) = 1$, and
- $((\{A_1, A_2, \ldots\} \subset \mathcal{A}) \wedge (i \neq j \Rightarrow A_i \cap A_j = \emptyset))$
  $\Rightarrow (P(\cup_{i=1}^{\infty} A_i) = \sum_{i=1}^{\infty} P(A_i))$.

Finally we define a probability space.

Definition 1.3 Let Ω be a set, $\mathcal{A}$ be a σ-algebra of subsets of Ω and $P : \mathcal{A} \to [0, 1]$ be a probability measure. Then we say that

- $(\Omega, \mathcal{A}, P)$ is a probability space,
- the set Ω is called the sample space, and,
- elements of $\mathcal{A}$ are called events.

The next definition sets the stage for exploring relations between events.

Definition 1.4 Let $(\Omega, \mathcal{A}, P)$ be a probability space, and let A and B be events. We say that A and B are independent events if

$$P(A \cap B) = P(A)P(B).$$

1.2 Problems

(1) Suppose that A, B, C are 3 distinct subsets of Ω. We can construct other distinct subsets of Ω from these 3 subsets using the only the operations $\cap$ and $\cup$ repeatedly: $A, B, C, A \cap B, (A \cap B) \cup C$ etc. We say that 2 subsets are different if they are not necessarily equal to each other. For example, $A \cap B$ is different from $A \cap C$, but $(A \cap B) \cup C$ is not different from $(A \cup C) \cap (B \cup C)$.
According to Rényi [54, p.26] there are 18 different subsets that can be constructed in this way having started with $n = 3$ distinct subsets A, B, C. Indeed,

- if $n = 2$, we could create 4 different subsets;
- if $n = 3$, we could create 18 different subsets;
- if $n = 4$, we could create 166 different subsets;
- if $n = 5$, we could create 7,579 different subsets;
- if $n = 6$, we could create 7,828,352 different subsets.

This leads to the sequence: $\{4, 18, 166, 7579, 7828352, \ldots\}$.
Guess the next number in the sequence.
Hint: Explore this sequence on the site

`<www.research.att.com/~njas/sequences>`

(2) Let A, B, C, D be subsets of Ω. Prove the following.

(a) $((A \cap B) \cup (C \cap D))' = (A' \cup B') \cap (C' \cup D')$
(b) $(A \cup B) \cap (A \cup B') \cap (A' \cup B) \cap (A' \cup B') = \emptyset$
(c) $A \triangle \Omega = A'$
(d) $A \setminus B = A \cap (A \triangle B)$
(e) $A \cup B = (A \triangle B) \triangle (A \cap B)$
(f) $A \triangle (B \triangle C) = (A \triangle B) \triangle C$ — another associative law.
Hint: Show that each side equals

$$(A \cap B \cap C) \cup (A \cap B' \cap C') \cup (A' \cap B \cap C') \cup (A' \cap B' \cap C).$$

(g) $(A \cap B') \triangle (B \cap A') = A \triangle B$
(h) $A \triangle B = C \triangle D \Rightarrow A \triangle C = B \triangle D$
(i) $A \cap (B \triangle C) = (A \cap B) \triangle (A \cap C)$
(j) $(A \triangle B) = (A \triangle C) \triangle (C \triangle B)$

(3) Let $A \subset \Omega$. Define, I_A, the indicator function (sometimes called the characteristic function) of A by

$$I_A : \Omega \to [0, 1]$$

where

$$I_A(x) := \begin{cases} 1 & \text{if } x \in A \\ 0 & \text{if } x \in A'. \end{cases}$$

(a) Let A, B be subsets of Ω. Show that

$$A = B \text{ iff } I_A = I_B.$$

(b) Prove the following identitites.

i. $I_\Omega = 1$; $I_\emptyset = 0$.
ii. $I_{A \cap B} = I_A I_B$
iii. $I_{A \cup B} = I_A + I_B - I_A I_B$
iv. $I_{A'} = 1 - I_A$
v. $I_{A \triangle B} \equiv I_A + I_B \pmod 2$
vi. $I_{A \setminus B} = I_A(1 - I_B)$

(c) Using the identitites concerning indicator functions in question 3b, prove the identitites concerning sets in question 2.

(4) Let Ω be a set and suppose that $\mathcal{R}$ is a non-empty set of subsets of Ω. We say that $\mathcal{R}$ is a **ring** of subsets of Ω if

$$(A \in \mathcal{R} \text{ and } B \in \mathcal{R}) \Rightarrow (A \cup B \in \mathcal{R} \text{ and } A \setminus B \in \mathcal{R}).$$

(a) Suppose that $\mathcal{R}$ is a ring of subsets of Ω. Prove that $\emptyset \in \mathcal{R}$.
(b) Give an example of a ring $\mathcal{R}$ of subsets of Ω such that $\Omega \notin \mathcal{R}$.
(c) Let $\mathcal{R}$ be a set of subsets of Ω. Prove that $\mathcal{R}$ is a ring if and only if

$$(A \in \mathcal{R} \text{ and } B \in \mathcal{R}) \Rightarrow (A \cap B \in \mathcal{R} \text{ and } A \triangle B \in \mathcal{R}).$$

(Hint: Use Problems 2d and 2e.)
(d) Let $\mathcal{S}$ be a set of subsets of Ω. Suppose that

$$(A \in \mathcal{S} \text{ and } B \in \mathcal{S}) \Rightarrow (A \cap B \in \mathcal{S} \text{ and } A \setminus B \in \mathcal{S}).$$

Show that $\mathcal{S}$ is **not necessarily** a ring of subsets of Ω.
(e) Show that the intersection of two rings of subsets of Ω is a ring of subsets of Ω.

(Source: Halmos [29, §4]. This is one of the great books in mathematics.)

(5) Write a 500—750 word essay on *The Life and Work of A.N. Kolmogorov.*

(6) Let $(\Omega, \mathcal{A}, P)$ be a probability space. Prove the following statements.

(a) $\emptyset \in \mathcal{A}$.
(b) $P(\emptyset) = 0$.
(c) $A \in \mathcal{A} \Rightarrow P(A') = 1 - P(A)$
(d) $((A \in \mathcal{A}) \wedge (B \in \mathcal{A})) \Rightarrow A \cap B \in \mathcal{A}$.
(e) $((A \in \mathcal{A}) \wedge (B \in \mathcal{A})) \Rightarrow P(A \cup B) = P(A) + P(B) - P(A \cap B)$.

(7) Let $(\Omega, \mathcal{A}, P)$ be a probability space.

(a) Let A, B, C be events. Prove that

$$\begin{aligned} P(A \cup B \cup C) &= P(A) + P(B) + P(C) \\ &\quad - P(A \cap B) - P(A \cap C) - P(B \cap C) \\ &\quad + P(A \cap B \cap C). \end{aligned} \tag{1.1}$$

(b) Let A, B, C, D be events. Write down an expression for $P(A \cup B \cup C \cup D)$ which is analogous to equation (1.1).
(c) Let $A_1, A_2, \ldots, A_n$ be events. Find an expression for $P(A_1 \cup \ldots \cup A_n)$ which generalises (1.1). (This is a challenge in mathematical notation!)

(8) Let $(\Omega, \mathcal{A}, P)$ be a probability space.

(a) Let A, B be events. Prove that

$$P(A \cup B) \leq P(A) + P(B). \tag{1.2}$$

Inequality (1.2) is known as Boole's inequality.
(b) Hence derive a generalisation of (1.2) for 3 events A_1, A_2, A_3.
(c) State and prove a generalisation of (1.2) for n events

$$A_1, A_2, \ldots, A_n.$$

(9) Let $(\Omega, \mathcal{A}, P)$ be a probability space.

(a) Let A, B be events. Prove that

$$P(A \cap B) \geq P(A) + P(B) - 1. \tag{1.3}$$

This is an instance of Bonferroni's inequality.
(b) Using the fact that

$$A \cap B \cap C = (A \cap B) \cap C,$$

derive a generalisation of (1.3) for 3 events A_1, A_2, A_3.
(c) Hence prove the general inequality:

$$P(\cap_{i=1}^n A_i) \geq \sum_{i=1}^n P(A_i) - (n-1).$$

(10) Let $(\Omega, \mathcal{A}, P)$ be a probability space.

(a) Let A, B be events. Prove that

$$P(A \cup B) \geq 1 - P(A') - P(B'). \tag{1.4}$$

(b) Hence derive a generalisation of (1.4) for 3 events A_1, A_2, A_3.

(c) State and prove a generalisation of (1.4) for n events

$$A_1, A_2, \ldots, A_n.$$

(Hint: Use induction.)

(11) Let $(\Omega, \mathcal{A}, P)$ be a probability space. If A, B are any two events, we define the distance $d(A, B)$ between A and B by

$$d(A, B) = P(A \triangle B).$$

Prove that if A, B, C are events then

$$d(A, C) \leq d(A, B) + d(B, C).$$

This is the triangle inequality for the distance function d. (Hint: Problem 2j.)

(12) Let $(\Omega, \mathcal{A}, P)$ be a probability space. If A, B are any two events, we define the distance $d(A, B)$ between A and B by

$$d(A, B) = \begin{cases} \dfrac{P(A \triangle B)}{P(A \cup B)} & \text{if } P(A \cup B) \neq 0, \\ 0 & \text{if } P(A \cup B) = 0. \end{cases}$$

Prove that if A, B, C are events then

$$d(A, C) \leq d(A, B) + d(B, C).$$

This is the triangle inequality for the distance function d.

(13) Let Ω be the set of real numbers and let $\mathcal{A}$ be the set of intervals defined by

$$\mathcal{A} := \{(x, \infty) : x \in \Omega\}.$$

Prove that the set $\mathcal{A}$ is closed under the operations of union and intersection, but not under the operation of complementation. (Source: Stoyanov [64, p. 73] is a very interesting source of detailed information on counterexamples in probability.)

(14) Let A, B be subsets of Ω. Find necessary and sufficient conditions which ensure that there is a set X such that $X \subset \Omega$ and

$$(A \cap X) \cup (B \cap X') = \emptyset.$$

This is an exercise in solving equations where the unknowns are sets. (Source: Stoyanov, Mirazchiiski, Ignatov and Tanushev [65, p. 73, Exercise 11.5]).

(15) Let $(\Omega, \mathcal{A}, P)$ be a probability space. Suppose that A, B are events and $A \cap B = \emptyset$. Prove that $P(A) \leq P(B')$.

(16) Let $(\Omega, \mathcal{A}, P)$ be a probability space.

(a) Let $A_1, A_2, A_3, \ldots$ be an infinite sequence of events such that

$$A_1 \subset A_2 \subset A_3 \subset \ldots$$

and define

$$A = \cup_{n=1}^{\infty} A_n.$$

Prove that

$$\lim_{n \to \infty} P(A_n) = P(A).$$

(b) Let $A_1, A_2, A_3, \ldots$ be an infinite sequence of events such that

$$A_1 \supset A_2 \supset A_3 \supset \ldots$$

and define

$$A = \cap_{n=1}^{\infty} A_n.$$

Prove that

$$\lim_{n \to \infty} P(A_n) = P(A).$$

(c) Let $A_1, A_2, A_3, \ldots$ be an infinite sequence of events. Define

$$\limsup A_n = \cap_{j=1}^{\infty} \cup_{i=j}^{\infty} A_i$$

and

$$\liminf A_n = \cup_{j=1}^{\infty} \cap_{i=j}^{\infty} A_i$$

i. Prove that $x \in \limsup A_n$ if and only if x is an element of infinitely many of the events $A_1, A_2, A_3, \ldots$.

ii. Prove that $x \in \liminf A_n$ if and only if x is an element of all but finitely many of the events $A_1, A_2, A_3, \ldots$.

iii. Prove that $\liminf A_n \subset \limsup A_n$.

iv. Prove that $\liminf(A_n') = (\limsup A_n)'$.

v. Prove that

$$(\limsup A_n) \setminus (\liminf A_n) = \limsup(A_n \triangle A_{n+1}).$$

vi. Prove that

$$\limsup_{n\to\infty} P(A_n) \leq P(\limsup A_n)$$

and

$$\liminf_{n\to\infty} P(A_n) \geq P(\liminf A_n).$$

(17) **Conditional probability.** For many students, problems about conditional probability appear to be tricky. The aim of this exercise is to place this concept on a solid mathematical foundation. Let $(\Omega, \mathcal{A}, P)$ be a probability space. Let $B \in \mathcal{A}$ be an event such that $P(B) > 0$. Define the function

$$\begin{array}{rccl} P_B: & \mathcal{A} & \to & \Re \\ & A & \mapsto & P_B(A) := P(A \cap B)/P(B)\,. \end{array}$$

(a) Prove that $(\Omega, \mathcal{A}, P_B)$ is a probability space. We usually write $P_B(A)$ as $P(A|B)$.

(b) **Bayes' theorem.** Suppose that

- $\{X, Y_1, Y_2, \ldots, Y_n\} \subset \mathcal{A}$;
- $\Omega = Y_1 \cup Y_2 \cup \ldots \cup Y_n$;
- if $i \neq j$ then $Y_i \cap Y_j = \emptyset$.

Prove that

$$P(X) = \sum_{i=1}^{n} P(Y_i)P(X|Y_i)\,.$$

Hence prove Bayes' theorem:

$$P(Y_k|X) = \frac{P(Y_k)P(X|Y_k)}{\sum_{i=1}^{n} P(Y_i)P(X|Y_i)} \quad (1 \leq k \leq n).$$

[Hint: $P(Y_k|X) = P(Y_k \cap X)/P(X)$.]

(18) **Application of conditional probability.** The following example is based on material found in the delightful book by Deborah Bennett called *Randomness* [7, pp.2–3] which, in turn, is based on work by D. Kahneman and A. Tversky [69, pp. 156–158].

A taxi was involved in a hit and run accident at night. There are two taxi companies in the city, namely Black Taxis and White Taxis. We know that 85% of the taxis in the city are Black and 15% are White. There was a witness to the accident and, according to the witness, the taxi involved in the accident was White. Now further investigation of the reliability of the witness showed that, under similar conditions, the witness was able to identify correctly the colour of a taxi 80% of the time.

(a) Without any calculation, do you think it is more likely that the taxi involved was Black or White?
(b) Calculate the probability that the taxi involved was White.
(c) Compare your answers to these two questions.
(d) Explore the sensitivity of the answer to Question 18b to the data as follows. Suppose that $0 \leq p \leq 1$ and that $100p\%$ of the taxis are White and $100(1-p)\%$ of the taxis are Black. Leave the reliability of the witness at 80%. Show that the probability that the taxi involved is White, given that the witness claims that the taxi involved was White, exceeds 0.5 if and only if $p > 0.2$.
(e) Explore the sensitivity of the answer to Question 18b to the data even further. Suppose that $0 \leq p \leq 1$ and that $100p\%$ of the taxis are White and $100(1-p)\%$ of the taxis are Black. Suppose that $0 \leq q \leq 1$ and that the reliability of the witness is 100q%; i.e. the witness is able to identify correctly the colour of a taxi 100q% of the time. Determine the region inside the square

$$\{(p, q) : 0 \leq p \leq 1, 0 \leq q \leq 1\}$$

of points for which the probability that the taxi involved is White, given that the witness claims that the taxi involved was White, exceeds 0.5.

Chapter 2

Elementary probability

2.1 Notes

This section deals with "elementary" probability because we consider the simplest of all probability spaces which we now define. Let $(\Omega, \mathcal{A}, P)$ be a probability space in which

- Ω is a finite set,
- $\mathcal{A}$ is the power set of Ω (i.e. the set of all subsets of Ω), and,
- $P : \mathcal{A} \to [0, 1]$ is defined by $P(A) = \dfrac{\#A}{\#\Omega}$ (where $\#X$ denotes the number of elements in a set X).

Thus, in the sample space Ω, all points are equally likely to occur.

Before going on to the problems you ought to check that we have a well-defined probability space. Although this probability space is extremely simple, we see in the problems below, that it offers many challenges and many opportunities for applications.

2.2 Problems

(1) A survey was conducted of a random sample of 100 persons who hold a particular credit card about their intentions to travel abroad within the next 12 months. Each person was asked the question *Are you planning to travel abroad next year?* and their answer was recorded as *Yes*, *Undecided*, or *No*. The age of the respondent was also recorded. The data were analysed in terms of the age groups of the respondents and the results can be found in Table 2.1.

Table 2.1 Travel survey data

		Age group			
		25 or less	26–40	41 or more	Total
Response	*Yes*	2	12	15	29
	Undecided	5	10	16	31
	No	10	15	15	40
	Total	17	37	46	100

Suppose that we choose a card holder at random from the set of all card holders. Use the data in Table 2.1 to estimate the probabilities of each of the following events:

(a) that the card holder **either** intends to travel in the next 12 months **or** is undecided;
(b) that the card holder's age is 40 or less;
(c) that the card holder's age is 25 or less **and** the card holder intends to travel abroad within the next 12 months;
(d) that the card holder's age is 25 or less **or** the card holder intends to travel abroad within the next 12 months;
(e) that, **assuming that we know that the card holder's age is 25 or less**, the card holder intends to travel abroad within the next 12 months;
(f) that, **assuming that we know that the card holder intends to travel abroad within the next 12 months**, the card holder's age is 25 or less;
(g) that the card holder's age does not exceed 40 **and** the card holder is not undecided about travelling abroad next year;
(h) that the card holder's age does not exceed 40 **or** the card holder is not undecided about travelling abroad next year.

(2) The last 3 digits of a Bendigo telephone number have been erased and all that we know is that the number was 54426???. Assuming that all possibilities are equally likely, find the probabilities of the following events.

(a) The missing digits are 7, 4, 0 (in this order).
(b) The set of missing digits is $\{0, 4, 7\}$.

(c) The missing digits are all equal to each other.
(d) Two of the missing digits are equal to each other, but the third is different from the other two.
(e) All three digits are different from each other.
(f) You can check that the answers to 2c, 2d, and 2e should sum to 1.00: why is it so?

(3) The last digit of a telephone number chosen at random from your local telephone directory could be 0,1,2,. . .,8 or 9. By taking a random sample of 200 numbers, does it appear that all possibilities are equally likely? (This can be tested formally using a χ^2 goodness-of-fit test.)
(4) In the state of Victoria in Australia, Tattslotto consists of choosing numbers at random from the set $\{1, 2, 3, \ldots, 45\}$. Up until 19 September 1998, the the 45 numbers have been chosen with the frequencies shown in Table 2.2. Thus, in all draws, the number 21 has been drawn on 115 occasions. (Source: [67])

Table 2.2 Tattslotto frequencies

1 (130)	2 (126)	3 (138)	4 (126)	5 (125)
6 (106)	7 (131)	8 (146)	9 (129)	10 (109)
11 (131)	12 (121)	13 (124)	14 (122)	15 (132)
16 (120)	17 (121)	18 (121)	19 (135)	20 (123)
21 (115)	22 (112)	23 (135)	24 (110)	25 (130)
26 (127)	27 (103)	28 (119)	29 (120)	30 (117)
31 (126)	32 (113)	33 (123)	34 (117)	35 (130)
36 (124)	37 (111)	38 (110)	39 (101)	40 (128)
41 (128)	42 (130)	43 (125)	44 (129)	45 (118)

(a) Represent the data in Table 2.2 in a frequency histogram.
(b) Does it appear that all numbers have an equal chance of being drawn? You should test this formally with a χ^2 goodness-of-fit test.

(5) In the state of Victoria, Tattslotto consists of choosing six numbers at random and without replacement from the set $\{1, 2, 3, \ldots, 45\}$. Without calculation, compare the probabilities of each of the fol-

lowing events.

(a) That the winning numbers this week will be

$$\{1, 2, 3, 4, 5, 6\}.$$

(b) That the winning numbers this week will be

$$\{11, 20, 22, 31, 35, 37\}.$$

(c) That the winning numbers this week will be

$$\{11, 20, 22, 31, 35, 37\}$$

given that the winning numbers last week were also

$$\{11, 20, 22, 31, 35, 37\}.$$

(Source: Lotteries provide research questions for those interested in studying either mathematics or human behaviour: see N. Henze and H. Riedwyl [32] and R. Clark [15], [16].)

(6) A number X is chosen at random from the set $\{1, 2, \ldots, n\}$. Let $p(n)$ denote the probability that $X^2 - 1$ is divisible by 10. Find

(a) $p(10)$
(b) $p(25)$
(c) $\lim_{n\to\infty} p(n)$.

(Source: Based on B. Sevastanyov , V. Christyakov and A. Zubkov [59, p. 14].)

(7) Consider the following three games.

Game 1: Throw a single die 6 times. If at least once, your score is 1 then you win the game.
Game 2: Throw a single die 12 times. If at least twice, your score is 1 then you win the game.
Game 3: Throw a single die 18 times. If at least thrice, your score is 1 then you win the game.

In which game do you have the highest probability of winning? (Source: Historical notes on this problem can be found in Mosteller [47, Problem 19] which is a classic work on problems in elementary probability.)

(8) Suppose that we have 4 boxes A, B, C, D.

- Box A contains 6 white balls, 3 red balls and 1 black ball.
- Box B contains 4 white balls and 6 red balls.
- Box C contains 8 white balls and 2 red balls.
- Box D is empty.

We will place one ball in D according to the following procedure.

- Choose a ball at random from A.
- If the ball chosen from A is white, then choose a ball at random from B and place it in D.
- If the ball chosen from A is red, then choose a ball at random from C and place it in D.
- If the ball chosen from A is black, then choose one of the boxes B or C at random, choose a ball at random from that box and place it in D.

(a) Represent the above procedure by a tree diagram.
(b) What is the probability that the first ball chosen from A was white?
(c) What is the probability that the ball placed in D is red?
(d) Suppose that we know that the ball placed in D was red. What is the probability that the first ball chosen from A was white?
(e) Compare your answers to 8b and 8d.

(9) Suppose that there are n people in a room and that the birthdays of these people were randomly chosen from the 365 days of the year. Let $p(n)$ denote the probability that there are at least 2 persons in the room with their birthdays on the same day during the year.

(a) Find an expression for $p(n)$ as a function of n. [Hint: Calculate the probability that all n birthdays are different.]
(b) Sketch the graph of $p(n)$ as a function of n.
(c) Show that if $n \geq 23$ then $p(n) > 0.5$.
(d) Give a non-technical interpretation of your answer to 9c.
(e) Show that if $n \geq 41$ then $p(n) > 0.9$.

(10) Suppose that there are n people in a room and that the birthdays of these people were randomly chosen from the 365 days of the year. Let $q(n)$ denote the probability that there is at least one person in the room whose birthday is on 29 May.

(a) Find an expression for $q(n)$ as a function of n.
(b) Sketch the graph of $q(n)$ as a function of n.
(c) Solve the equation $q(n) = p$ for n in terms of p.
(d) Show that if $n \geq 253$ then $q(n) > 0.5$.
(e) Give a non-technical interpretation of your answer to 10d.
(f) For what values of n do we have $q(n) > 0.9$?

(11) Two players A, B are involved in a game. They take up to n turns in throwing 2 dice and A has first turn. Here are the rules of the game.

- If, in any turn, A's total score is 6 before B's total score on any turn is 7, then A wins the game.
- If, in any turn, B's total score is 7 before A's total score on any turn is 6, then B wins the game.
- If, after each player has had n turns, there is no winner then the game is declared to be a draw.

Let

- $p_A(n) =$ the probability that A is the winner,
- $p_B(n) =$ the probability that B is the winner, and
- $d(n) =$ the probability that the game is drawn.

(a) Find a formula for $p_A(n)$ as a function of n.
(b) Find a formula for $p_B(n)$ as a function of n.
(c) Find a formula for $d(n)$ as a function of n.
(d) Find $\lim_{n\to\infty} p_A(n)$, $\lim_{n\to\infty} p_B(n)$, and $\lim_{n\to\infty} d(n)$.

(Source: Uspensky [70, p.41])

(12) Table 2.3 is an extract from life tables for Australian males [6]. It depicts what would happen to a cohort of 100,000 Australian males all born in the same year given the current age-specific death rates among Australian males. (The number 100,000 is purely arbitrary: it is called the radix of the life table.) For each age x, $\ell(x)$ denotes the number of males in this cohort who survive to age x.

(a) Your first task is to find expressions for the following functions in terms of the function ℓ.

- $d(x) =$ the number of deaths in the year of age x to $x + 1$ among the $\ell(x)$ males who survive to age x,

- $p(x)$ = probability that a male who reaches age x survives to age $x + 1$,
- $q(x)$ = probability that a male who reaches age x does not survive to age $x + 1$.

(b) Your next task is to calculate the probabilities of the following events:

i. that an Australian male survives to age 30;
ii. that an Australian male survives to age 20 but does not reach age 60;
iii. that an Australian male, who has survived to age 20, does not survive to age 60;
iv. that an Australian male does not survive to age 55.

(c) Describe how one would go about finding data to construct a life table for a nation. (Hint: Start with $q(x)$; see [6, p. 7 *et seq.*])

(13) Using the life table for Australian males 1985–1987 (Table 2.3) and the life table for Australian females 1985–1987 (Table 2.4), plot the graph of the age specific death rates ($q(x)$) against age (x) for females and males on the same set of axes. Reflect on the difference between the lives of males and females in Australia. (Often in demography, one plots $\log_{10} q(x)$ rather than $q(x)$ to get a better picture. Try it.)

Table 2.3 Life Table for Australian Males 1985–1987

x	$\ell(x)$	x	$\ell(x)$	x	$\ell(x)$	x	$\ell(x)$	x	$\ell(x)$
0	100,000	20	97,956	40	95,218	60	84,782	80	36,671
1	98,970	21	97,798	41	95,046	61	83,539	81	33,284
2	98,896	22	97,643	42	94,858	62	82,190	82	29,926
3	98,842	23	97,492	43	94,654	63	80,730	83	26,629
4	98,795	24	97,346	44	94,430	64	79,155	84	23,427
5	98,757	25	97,207	45	94,183	65	77,460	85	20,356
6	98,727	26	97,073	46	93,909	66	75,638	86	17,450
7	98,701	27	96,942	47	93,605	67	73,686	87	14,746
8	98,677	28	96,813	48	93,268	68	71,599	88	12,278
9	98,654	29	96,685	49	92,893	69	69,372	89	10,071
10	98,630	30	96,559	50	92,475	70	67,003	90	8,139
11	98,605	31	96,435	51	92,009	71	64,492	91	6,483
12	98,579	32	96,310	52	91,490	72	61,840	92	5,094
13	98,550	33	96,186	53	90,914	73	59,051	93	3,953
14	98,520	34	96,061	54	90,274	74	56,132	94	3,033
15	98,485	35	95,935	55	89,566	75	53,095	95	2,300
16	98,436	36	95,804	56	88,782	76	49,952	96	1,724
17	98,360	37	95,669	57	87,917	77	46,718	97	1,275
18	98,250	38	95,528	58	86,965	78	43,413	98	931
19	98,111	39	95,378	59	85,922	79	40,056	99	671

(The table above is an extract from Office of the Australian Actuary, *Australian Life Tables, 1985–87*, Canberra: Australian Government Publishing Service, 1991. Commonwealth of Australia copyright reproduced by permission [6].)

Table 2.4 Life Table for Australian Females 1985–1987

x	$\ell(x)$	x	$\ell(x)$	x	$\ell(x)$	x	$\ell(x)$	x	$\ell(x)$
0	100,000	20	98,670	40	97,495	60	91,425	80	58,031
1	99,206	21	98,617	41	97,392	61	90,764	81	54,722
2	99,144	22	98,566	42	97,279	62	90,045	82	51,237
3	99,102	23	98,516	43	97,152	63	89,259	83	47,597
4	99,073	24	98,467	44	97,011	64	88,399	84	43,829
5	99,049	25	98,417	45	96,855	65	87,458	85	39,968
6	99,029	26	98,368	46	96,681	66	86,426	86	36,059
7	99,009	27	98,319	47	96,488	67	85,296	87	32,151
8	98,990	28	98,269	48	96,273	68	84,059	88	28,298
9	98,973	29	98,219	49	96,035	69	82,706	89	24,558
10	98,957	30	98,168	50	95,773	70	81,228	90	20,988
11	98,942	31	98,116	51	95,484	71	79,620	91	17,646
12	98,929	32	98,062	52	95,166	72	77,875	92	14,584
13	98,915	33	98,006	53	94,820	73	75,986	93	11,840
14	98,899	34	97,948	54	94,442	74	73,946	94	9,438
15	98,879	35	97,886	55	94,032	75	71,746	95	7,386
16	98,853	36	97,821	56	93,588	76	69,374	96	5,674
17	98,819	37	97,750	57	93,110	77	66,823	97	4,280
18	98,774	38	97,672	58	92,592	78	64,085	98	3,171
19	98,723	39	97,587	59	92,032	79	61,154	99	2,308

(The table above is an extract from Office of the Australian Actuary, *Australian Life Tables, 1985–87*, Canberra: Australian Government Publishing Service, 1991. Commonwealth of Australia copyright reproduced by permission [6].)

(14) **Poker.** Gambling provides many interesting problems in probability. See Epstein [23] for a thorough exposition. In fact, the study of games of chance was the motivation for the first books on probability; with the advent of casinos, tattslotto and internet gambling, such games are still providing new problems in probability (and new issues for society). In this exercise, we demonstrate this link by considering the game of Poker.

We assume that Poker is being played with a standard deck of 52 cards: 4 suits ($\heartsuit$, $\diamondsuit$, $\clubsuit$, and $\spadesuit$) and 13 face values in each suit (A, 2, 3, 4, 5, 6, 7, 8, 9, 10, J, Q, K). Each layer is dealt a hand of 5 cards which we assume to be chosen at random from the deck. We will say that 5 cards are in "sequence" if their face values are consecutive such as $\{3, 4, 5, 6, 7\}$. For this purpose, an Ace can count "low" $\{A, 2, 3, 4, 5\}$ or "high" $\{10, J, Q, K, A\}$.

(a) Prove that there are 2,598,960 possible Poker hands.

(b) Verify the results shown in Table 2.5.

Table 2.5 Poker Hands

Type of hand	Example	Number of such hands	Prob. of such a hand
straight flush: all 5 same suit & in sequence; Ace can be high or low.	9 ♣, 10 ♣, J ♣, Q ♣, K ♣	40	0.000015
4 of a kind: 4 cards of same face value;	9, 9, 9, 9, 2	624	0.000240
full house: 3 cards of same face value 2 cards with 1 other face value	8, 8, 8, 4, 4	3,744	0.001441
flush: 5 cards of same suit but not in sequence	2 ♠, 4 ♠, 7 ♠, Q ♠, K ♠	5,108	0.001965
straight: 5 cards in sequence but not of same suit	A ♠, 2 ♡, 3 ♡, 4 ♣, 5 ♠	10,200	0.003925
3 of a kind: 3 cards of same face value and 2 cards with 2 different values	9, 9, 9, K, 2	54,912	0.021128
2 pair: 2 cards of same face value and 2 cards with 1 different value, 1 card with 1 different value;	9, 9, K, K, 2	123,552	0.047539
1 pair: 2 cards of same face value and 3 cards with 3 different values;	9, 9, 7, K, 2	1,098,240	0.422569
none of the above		1,302,468	0.501150

Chapter 3

Discrete random variables

3.1 Notes

We begin with some general aspects of discrete random variables. Let us begin with the definition of a random variable (r.v.).

Definition 3.1 Let $(\Omega, \mathcal{A}, P)$ be a probability space. A random variable is a function

$$X : \Omega \to \Re$$

such that, for any $s \in \Re$, $X^{-1}((-\infty, s]) \in \mathcal{A}$.

We write $P(X \leq s) := P(X^{-1}((-\infty, s])) = P(\{\omega : X(\omega) \leq s\})$. In the definition, insisting that $X^{-1}((-\infty, s]) \in \mathcal{A}$ ensures that $P(X^{-1}((-\infty, s]))$ is well defined because the domain of P is $\mathcal{A}$.

Now we can define a discrete r.v.

Definition 3.2 We say that X is a discrete r.v. if the function X assumes only a finite or countable number of values.

In this chapter, all random variables are discrete.

Definition 3.3 The *probability function* for X is the function $P(X = t)$ (as a function of t). The key properties of this function are:

- $P(X = t) \geq 0$,
- $\sum_t P(X = t) = 1$.

The following definition introduces the concept of expected value.

Definition 3.4 Here we define a number of quantities which, in certain circumstances, may not exist.

The *expected value* of X is

$$E(X) = \mu_X := \sum_t tP(X = t).$$

Let $g : \Re \to \Re$ be a suitable function. Then the expected value of $g(X)$ is

$$E(g(X)) := \sum_t g(t)P(X = t).$$

The variance of X is $\text{var}\,(X) = \sigma_X^2 := E((X - \mu_X)^2)$.
The standard deviation of X is σ_X.
It is easy to prove that

$$\text{var}(X) = \sigma_X^2 = E(X^2) - E(X)^2.$$

The covariance of two random variables X and Y is

$$\text{Cov}(X, Y) = E(XY) - E(X)E(Y).$$

The coefficient of correlation of two random variables X and Y is

$$\text{Corr}(X, Y) = \frac{\text{Cov}\,(X, Y)}{\sigma_X \sigma_Y}.$$

The moment generating function (m.g.f.) of X is

$$M_X(t) := E(\exp(tX)) = \sum_{k=0}^{\infty} \frac{E(X^k)t^k}{k!}.$$

The characteristic function (c.f.) of X is

$$\phi_X(t) := E(\exp(itX)).$$

We now consider some specific discrete random variables. Many more discrete probability distributions can be found in the *magnum opus* by Wimmer and Altman [72].

Distribution	$P(X = t)$	Parameters
Uniform	$\frac{1}{b-a+1}$, $(t = a, a+1, \ldots, b)$	a, b
Binomial	$\binom{n}{t} p^t (1-p)^t$, $(t = 0, 1, \ldots, n)$	n, p
Poisson	$\frac{\exp(-\lambda)\lambda^t}{t!}$, $(t = 0, 1, 2, \ldots)$	λ
Hypergeometric	$\frac{\binom{M}{t}\binom{N-M}{n-t}}{\binom{N}{n}}$, $t_1 \leq t \leq t_2$ where $t_1 = \max(0, n - N - M)$ and $t_2 = \min(M, n)$	M, N, n
Geometric	$p(1-p)^t$, $(t = 0, 1, 2, \ldots)$	p

Note: $X \sim$ Poisson(λ) means "X is a random variable which has a Poisson distribution with parameters λ"; similar notation will be used for other distributions. Thus $\sim$ means "is distributed as".

3.2 Problems

(1) In probability theory, one often encounters a statement such as "Assume that X is a r.v. with $E(|X|^3) < \infty$". From statements such as this, it seems that we cannot be assured that any r.v. has a finite moment of a particular order. This example show us why such assumptions are necessary.
Find a discrete random variable X for which

(a) μ_X does not exist
(b) μ_X exists but σ_X does not exist
(c) $E(X^k)$ exists for $k = 0, 1, 2, \ldots, p$ but not for $k = p + 1, p + 2, \ldots$.

(Hint: $\sum_{t=1}^{\infty} t^{-s}$ converges if $s > 1$ and diverges if $s \leq 1$.)

(2) In any share portfolio it is important to understand the association between various investments. For example, it is helpful to know that, if one stock is not performing well, then we can expect that some other investment may be performing better. Such associations tend to reduce the risk of the overall investment. This example is designed to allow you to see the role that correlation plays in portfolio analysis.
Suppose that $X\%$ and $Y\%$ represent the possible returns in the next 12 months on two investments in a portfolio. Suppose that, according to the experts, Table 3.1 summarises all the possible values of X and Y and their probabilities. This table is the *joint probability distribution* of X and Y. Note that four entries are missing from the table: these are indicated by A, B, C and D.

(a) Find the four missing numbers in Table 3.1.
(b) Calculate the following probabilities.

i. $P(X = 0 \text{ and } Y = 2)$
ii. $P(X = 0 \text{ and } Y \leq 1.5)$
iii. $P(X = 0 \text{ or } Y = 3)$
iv. $P(X = 2)$
v. $P(X \neq 2 \text{ and } Y = -1)$
vi. $P(X < 0 \text{ or } Y \neq 1)$

Table 3.1 Values of $P(X = s, Y = t)$.

		s				
		-2	0	2	10	Total
t	-1	0.01	0.03	0.05	0.04	0.13
	0	0.04	0.06	A	0.04	0.23
	1	0.08	0.10	0.10	0.02	0.30
	2	0.05	B	0.03	0.01	0.20
	3	0.06	0.05	0.02	0.01	0.14
Total		0.24	C	0.29	D	1.00

(c) Find the expected value of XY:

$$E(XY) = \sum stP(X = s, Y = t).$$

(d) Find the distribution of X by completing the following table:

s	-2	0	2	10
$P(X = s)$				

(e) Find $E(X) = \mu_X$, $E(X^2)$ and σ_X^2.

(f) Find the distribution of Y by completing the following table.

t	-1	0	1	2	3
$P(Y = t)$					

(g) Find $E(Y) = \mu_Y$, $E(Y^2)$ and σ_Y^2.

(h) Find the coefficient of correlation between X and Y:

$$\rho_{X,Y} = \frac{E(XY) - E(X)E(Y)}{\sigma_X \sigma_Y}.$$

(i) Do the entries in Table 3.1 confirm that the coefficient of correlation between X and Y ought to be negative?

(j) Explain the advantage of having in a portfolio two investments with returns which are negatively correlated.

(3) Let X and Y be independent random variables each of which is uniformly distributed over the set

$$\{0, 1, 2, \ldots, 9\}.$$

Consider the random variable XY and write

$$XY = 10\xi_1 + \xi_2$$

where $\{\xi_1, \xi_2\} \subset \{0, 1, 2, \ldots, 9\}$.

(a) Construct a table which shows (ξ_1, ξ_2) for all possible values (X, Y).
(b) Use the table constructed in the previous part to construct a table which represents the distribution of the vector (ξ_1, ξ_2).
(c) What is the most likely value of the vector (ξ_1, ξ_2)?
(d) What is the probability that XY is a perfect square?
(e) Show that the random variables ξ_1 and ξ_2 are not independent.
(f) Find the distribution of ξ_1, $E(\xi_1)$, and $\mathrm{Var}(\xi_1)$.
(g) Find the distribution of ξ_2, $E(\xi_2)$, and $\mathrm{Var}(\xi_2)$.
(h) Find $E(\xi_1\xi_2)$, $\mathrm{Cov}(\xi_1, \xi_2)$, $\mathrm{Corr}(\xi_1, \xi_2)$.

(Source: This problem was inspired by [59, Problem 3.10].)

(4) This is an exercise in using discrete probability distributions in constructing mathematical models.

(a) Carefully describe a practical situation which would give rise to a random variable whose distribution you would expect to be a **binomial** distribution. State some sensible values of the parameters n and p involved.
(b) Carefully describe a practical situation which would give rise to a random variable whose distribution you would expect to be a **Poisson** distribution. State some sensible value of the parameter λ involved.
(c) Carefully describe a practical situation which would give rise to a random variable whose distribution you would expect to be a **uniform** distribution over the set $\{a, a+1, \ldots, b\}$. State some sensible values of the parameters a and b involved.
(d) Carefully describe a practical situation which would give rise to a random variable whose distribution you would expect to be a **hypergeometric** distribution. State some sensible values of the parameters N and n involved.

(5) This exercise is an introduction to Monte Carlo methods which are very popular in numerical analysis, physics and simulation. The

aim of this exercise is to use a stochastic approach to estimating the integral

$$I = \int_0^1 x^2\,dx.$$

(a) Draw the graph of the function $y = x^2$ for $0 \le x \le 1$ and enclose it in the square $ABCD$ where $A = (0,0)$, $B = (0,1)$, $C = (1,1)$ and $D = (1,0)$.

(b) Using a random number generator or table of random numbers, choose x at random from the interval $(0,1)$ and choose y at random from $(0,1)$: this leads to a point (x,y) chosen at random in the square $ABCD$.

(c) Repeat the process in 5b say $n = 100$ times and record the number (U) of points chosen which fall **under** the graph of $y = x^2$.

(d) Use $U/n = U/100$ as an estimate of the integral I.

(e) Explain the relevance of the binomial distribution to this problem.

(f) Use the Monte Carlo method to estimate the length of the arc of the graph

$$y = \sin x \quad (0 \le x \le \pi).$$

[Hint: Express the arc length as a definite integral. You will find the formula for arc length in standard calculus books. *En passant*, you might wonder why you cannot calculate this integral directly.]

(6) This exercise deals with a very famous inequality due to P.L. Chebyshev.
Let X be a discrete random variable with mean value μ_X and standard deviation σ_X.

(a) Prove Chebyshev's inequality:

$$P(|X| \ge c) \le c^{-2}E(X^2).$$

(b) Hence prove another form of Chebyshev's inequality

$$P(|X - \mu_X| \ge c) \le c^{-2}\sigma_X^2.$$

(c) Prove the following generalisation of Chebyshev's inequality. Let $\phi : (0, \infty) \to (0, \infty)$ be a monotonic increasing function and suppose that $E(\phi(|X|)) = M < \infty$. Prove that

$$P(|X| \geq c) \leq M/\phi(c).$$

(d) Prove that the result in 6c really generalises the result in 6a. (There is a huge literature on inequalities of the Chebyshev type. See, for example, the paper Savage [58] for a nice exposition of this field.)

(e) Mathematics is a human endeavour. It behoves us to learn something about those who made important contributions to our field. Write a short one page summary which sketches key aspects of the life and work of Chebyshev.

(Sources: Question 6c can be found in Feller [25, p.242, Problem 40] as can the the solutions to other parts of the question.)

(7) Suppose that the r.v. X has a Poisson distribution with parameter $\lambda > 0$.

(a) Show that $M_X(t) = \exp(\lambda(\exp(t) - 1))$.
(b) Hence, $E(X) = \lambda$, $E(X^2) = \lambda + \lambda^2$ and $\mathrm{Var}(X) = \lambda$.
(c) Show that

$$P(X = 0) \leq P(X = 1) \leq \ldots \leq P(X = [\lambda])$$

and

$$P(X = [\lambda]) \geq P(X = [\lambda] + 1) \geq P(X = [\lambda] + 2) \ldots.$$

(8) This problem shows us that the sum of independent Poisson r.vs. is a Poisson r.v.

(a) Suppose that $X_1 \sim \text{Poisson}(\lambda_1)$, $X_2 \sim \text{Poisson}(\lambda_2)$. Prove that $X_1 + X_2 \sim \text{Poisson}(\lambda_1 + \lambda_2)$.
(b) Extend the above result to a sum of n independent Poisson r.vs.

(9) A bank sells twenty thousand \$ 5 notes and subsequently discovers that 150 of these notes are counterfeit. The bank recovers 100 of the twenty thousand notes. Let X be the number of counterfeit notes among those recovered.

(a) Argue that $X \sim$ Binomial$(n = 100, p = 0.0075)$.
(b) Calculate $P(X \geq 2)$.
(c) Argue that $X \sim$ Poisson$(\lambda = 0.75)$ (approximately).
(d) Calculate $P(X \geq 2)$ approximately using the Poisson distribution.
(e) Discuss the Poisson approximation in terms of accuracy and ease.

(Source: Suggested by H. Tuckwell [68, p. 57].)

(10) Probability theory plays an important role in statistical quality control. A useful reference in this field is the book by A. Mitra [44]. The following problem is a classical problem in statistical quality control.
A firm manufactures tin cans. The cans are packed into very large batches and each batch is inspected by a quality control inspector before it leaves the factory according to the following sampling plan (which is a single sampling plan).
The inspector chooses a random sample of 15 cans from the batch and classifies each can either as conforming to specifications or as not conforming to specifications. If the number of non-conforming cans in the sample is 3 or more then the entire batch is rejected; otherwise the batch is accepted.

(a) Suppose that p is the proportion of cans in the process (and hence the batch) which are nonconforming. Let

$$Pa = \text{the probability that the batch is accepted}$$

as a function of p. Prove that

$$Pa = (1 - p)^{13}(1 + 13p + 91p^2).$$

(b) Draw the graph of Pa against p. This curve is known as the operating characteristic (O.C.) of the this single sampling plan.
(c) Estimate the value of p for which $Pa = 0.95$. (This value of p is known as the "acceptable quality level" because it represents the quality of a batch which will be very likely to pass the acceptance test.)

(11) Continuation of problem 10.

(a) In problem 10, how do we use the given information that the batches were "very large"?
(b) In the rest of this problem, we explore the complexities involved when the batch size is not so large.
Suppose that the batch size was 100 and $p = 0.05$. Calculate Pa. [Hint: Hypergeometric probabilities can be found in many spreadsheets.]
(c) Find the O.C. for the sampling plan and compare it with the O.C. found in 10b.
(d) Show that, for this sampling plan, the AQL lies in the interval $[0.06, 0.07]$.

(12) Variation of problem 10.
A firm manufactures tin cans. The cans are packed into very large batches and each batch is inspected by a quality control inspector before it leaves the factory according to the following sampling plan (which is a double sampling plan).
The inspector chooses a random sample of 15 cans from the batch and classifies each can either as conforming to specifications or as not conforming to specifications.

- If the number of non-conforming cans in the sample is 3 or more then the entire batch is rejected.
- If the number of non-conforming cans in the sample is 0 then the batch is accepted.
- If the number of non-conforming cans in the sample is 1 or 2 then a second random sample of 5 items is drawn. If the total number of non-conforming items in the two samples is 2 or less then the batch is accepted; otherwise the batch is rejected.

(a) Represent this double sampling plan with a tree-diagram.
(b) Suppose that p is the proportion of cans in the process (and hence the batch) which are nonconforming. Let us regard

$$Pa = \text{the probability that the batch is accepted}$$

as a function of p. Prove that

$$Pa = (1-p)^{15} + 15p(1-p)^{19} + 180p^2(1-p)^{18}.$$

(c) Draw the graph of Pa against p. This curve is known as the operating characteristic (OC) of this double sampling plan. Compare the OC for this double sampling plan with the OC found in 10b.

(d) Estimate the value of p for which $Pa = 0.95$.

(e) Estimate the AQL for this double sampling plan and compare it with the AQL from the single sampling plan in problem 10c.

(13) This question shows the relationship between the mathematics of discrete random variables and use of tables of random numbers.

(a) Suppose that X is a discrete r.v. which is uniformly distributed over the set $\{0, 1, 2, \ldots, N-1\}$. Suppose that A and B are constants. Prove that $AX + B$ is uniformly distributed over the set $\{B, A+B, 2A+B, \ldots, (N-1)A+B\}$.

(b) Show how you can use a table of random digits to select a random sample of 10 pages from a book which has 250 pages.

(14) Consider the following experiment: throw a single die. The experiment results in "success" if the die shows 4, 5 or 6 when it lands. Otherwise it is a failure. Repeat the experiment $n = 12$ times and record $X =$ number of successes.

(a) Find the probability function of this discrete r.v. X.

(b) A famous data set is known as Weldon's data which resulted from 4096 observations of X. It is presented in Table 3.2 below. Compare Weldon's data with your findings concerning the probability function of X. (Source: This data comes from [66, p. 23].)

(15) Table 3.3 below presents a frequency table for

$$X = \text{ the number of strikes which commence in a week}$$

over the period 1948–1959 in the UK. Thus, in this period there were 626 weeks; in 252 of these weeks, no strikes commenced but in 229 of these weeks, 1 strike began etc. (Source: [66, p.169].)

(a) Show that the mean value of X is about 0.90.

(b) Try to model the distribution of X using a Poisson distribution with $\lambda = 0.90$.

Table 3.2 Weldon's data.

X	0	1	2	3	4	5	6
Frequency	0	7	60	198	430	731	948
X	7	8	9	10	11	12	
Frequency	847	536	257	71	11	0	

Table 3.3 No. major strikes commencing in a week. (UK, 1948–1959.)

X	0	1	2	3	4+	Total no. weeks
Frequency	252	229	109	28	8	626

Chapter 4

Continuous random variables

4.1 Notes

The problems in this chapter will strengthen your skills in calculus. At the end of a course on calculus, students should be good at algebra; at the end of a course on probability, students should be good at calculus.

We begin with some general aspects of continuous random variables.

Definition 4.1 The *distribution function* for X is the function

$$F_X(t) := P(X \leq t)$$

(as a function of t).

Note the following properties of F_X.

- $F_X : \Re \to [0, 1]$.
- $F_X(t)$ is a non-decreasing function of t.
- $\lim_{t\to\infty} F_X(t) = 1$.
- $\lim_{t\to-\infty} F_X(t) = 0$.

Definition 4.2 The *median* of X is a number $\tilde{\mu}$ such that, for any number $\epsilon > 0$,

$$F_X(\tilde{\mu} - \epsilon) \leq 0.5 \leq F_X(\tilde{\mu} + \epsilon). \tag{4.1}$$

In all situations in this chapter, F_X is continuous and hence we can write simply that

$$F_X(\tilde{\mu}) = 0.5.$$

Note that, in the definition, we say that the "median of X is **a** number ..."; here we are acknowledging the possibility that there may be several numbers which satisfy the criterion in the equation (4.1).

Definition 4.3 We say that X is a *continuous* r.v. if the d.f. can be written as

$$F_X(t) = \int_{-\infty}^{t} f_X(u)\,du \quad (t \in \Re)$$

for some integrable function $f_X : \Re \to \Re$.

Definition 4.4 The *probability density function* (pdf) for X is the function

$$f_X(t) := \frac{dF_X(t)}{dt}.$$

It is very convenient from a notational point of view that the functions F_X and f_X have $\Re$ as their domain. Throughout this work, we use the standard definition that for $a \in \Re$,

$$\int_{a}^{\infty} \phi(u)du := \lim_{x\to\infty} \int_{a}^{x} \phi(u)du,$$

$$\int_{-\infty}^{a} \phi(u)du := \lim_{x\to-\infty} \int_{x}^{a} \phi(u)du$$

and

$$\int_{-\infty}^{\infty} \phi(u)du := \int_{-\infty}^{a} \phi(u)du + \int_{a}^{\infty} \phi(u)du.$$

Note the following properties of f_X.

- $(\forall t \in \Re)(f_X(t) \geq 0)$.
- $\displaystyle\int_{-\infty}^{\infty} f_X(t)\,dt = 1.$

We now introduce the concept of expected value for a continuous random variable.

Definition 4.5 Here we define a number of quantities as integrals which, in certain circumstances, may not converge or exist.

The *expected value* of X is

$$E(X) = \mu_X := \int_{-\infty}^{\infty} t f_X(t)\, dt.$$

Let $g : \Re \to \Re$ be a suitable function. Then the expected value of $g(X)$ is

$$E(g(X)) := \int_{-\infty}^{\infty} g(t) f_X(t)\, dt.$$

The variance of X is $\mathrm{var}\,(X) := E((X - \mu_X)^2)$.
It is easy to prove the very useful fact that

$$\mathrm{var}\ (X) = E(X^2) - E(X)^2.$$

The covariance of two random variables X and Y is

$$\mathrm{Cov}\ (X,Y) = E(XY) - E(X)E(Y).$$

The coefficient of correlation of two random variables X and Y is

$$\mathrm{Corr}\ (X,Y) = \frac{\mathrm{Cov}\ (X,Y)}{\sigma_X \sigma_Y}.$$

The moment generating function (m.g.f.) of X is $M_X(t) := E(\exp(tX))$.
The characteristic function (c.f.) of X is $\phi_X(t) := E(\exp(itX))$.

We now consider some specific continuous random variables.

Distribution	pdf $f_X(t)$	Parameters
Uniform	$\begin{cases} \frac{1}{b-a}, & (a \le t \le b) \\ 0, & (\text{elsewhere}) \end{cases}$	a, b
Normal	$\frac{1}{\sigma\sqrt{2\pi}} \exp\left(-\frac{(t-\mu)^2}{2\sigma^2}\right), \quad (t \in \Re)$	μ, σ
Cauchy	$\frac{1}{\pi(1+t^2)}, \quad (t \in \Re)$	
Gamma	$\begin{cases} \frac{1}{\Gamma(\alpha)\beta^\alpha} \exp\left(\frac{-t}{\beta}\right) t^{\alpha-1}, & (t > 0) \\ 0, & (t \le 0) \end{cases}$	α, β

Notation.

(1) We will write $X \sim N(\mu, \sigma^2)$ to mean that "X is a random variable which has a Normal distribution with parameters μ and σ"; similar notation will be used for other distributions. Thus "$\sim$" means "is distributed as".

(2) If $Z \sim N(0, 1)$ we say that Z has a standard Normal distribution.

(3) I tend to use upper case N when writing about the Normal distribution.

(4) If $Z \sim N(0,1)$ then the p.d.f. and d.f. of Z are denoted, respectively by

$$\begin{aligned}\phi(t) &:= \frac{1}{\sqrt{2\pi}}\exp(-t^2/2) \quad (t \in \Re) \\ \Phi(t) &:= \int_{-\infty}^{t}\phi(x)dx = \frac{1}{\sqrt{2\pi}}\int_{-\infty}^{t}\exp(-x^2/2)dx \quad (t \in \Re).\end{aligned}$$

4.2 Problems

(1) Find a continuous random variable X for which

(a) μ_X does not exist
(b) μ_X exists but σ_X does not exist
(c) $E(X^k)$ exists for $k = 0, 1, 2, \ldots, p$ but not for $k > p$.

(2) A generalisation of the factorial function is the gamma function $\Gamma : [0, \infty) \to [0, \infty)$ which is defined by

$$\Gamma(\alpha) := \int_0^{\infty} \exp(-t)t^{\alpha-1}\,dt \quad (\alpha > 0). \tag{4.2}$$

This function is very important in probability.

(a) The integral in equation (4.2) is an improper integral on two counts. First, the range of integration is infinite, and second, there may be some difficulty with the integrand at $t = 0$ depending on the value of α. So your first task is to prove that $\Gamma(\alpha)$ is well defined for all $\alpha > 0$.
(b) By induction, prove that

$$\Gamma(n) = (n-1)! \quad (n = 1, 2, 3, \ldots).$$

This shows that the gamma function extends the factorial function.
(c) The gamma function leads naturally to the gamma distribution. Suppose that X has a gamma distribution with param-

eters α and β. Then

$$f_X(t) = \begin{cases} \dfrac{1}{\Gamma(\alpha)\beta^\alpha} \exp\left(\dfrac{-t}{\beta}\right) t^{\alpha-1}, & (t > 0) \\ 0, & \text{(elsewhere)} \end{cases}$$

Using the above properties of the gamma function, verify that f_X is a well-defined p.d.f. and prove that

$$E(X^k) = \frac{\beta^k \Gamma(\alpha + k)}{\Gamma(\alpha)} \quad (k = 0, 1, 2, 3, \ldots).$$

(3) This exercise deals with the Normal distribution which is perhaps the best known and most important continuous distribution. The book by Patel and Read [50] contains a wealth of information about the Normal distribution. In fact, there are so many aspects of this distribution, you could study the Normal distribution for your entire life!
Let $X \sim N(\mu, \sigma^2)$.

(a) Prove that the function f_X is unimodal with mean, median and mode all equal to μ.
(b) Prove that the graph of $f_X(t)$ has points of inflexion at $t = \mu \pm \sigma$.
(c) Prove that $E((X - \mu)^4) = 3\sigma^4$. (Hint: Here you will need the previous result about the gamma function.)
(d) Examine the definitions of the Normal distribution in some elementary textbooks on statistics and comment on these definitions.
(e) Engineers often encounter the "error function" [53, p.254] which is defined by

$$\text{erf}\,(x) = \frac{2}{\sqrt{\pi}} \int_0^x \exp(-t^2)\,dt.$$

Explain the relationship between the error function and the Normal distribution.
(f) Let $Z = (X - \mu)/\sigma$. By considering

$$F_Z(t) = P(Z \leq t),$$

show that $Z \sim N(0,1)$. We refer to Z as the standard Normal distribution; its distribution function $F_Z = \Phi$ is tabulated in most statistics books.

(4) This is an exercise in using continuous probability distributions in constructing mathematical models.

(a) Carefully describe a practical situation which would give rise to a random variable whose distribution you would expect to be a **uniform** distribution over the set (a, b). State some sensible values of the parameters a and b involved.

(b) Carefully describe a practical situation which would give rise to a random variable whose distribution you would expect to be a **Normal** distribution. State some sensible values of the parameters μ and σ involved.

(c) Carefully describe a practical situation which would give rise to a random variable whose distribution you would expect to be a non-symmetric or skewed distribution.

(5) In many elementary statistics books, we find a statement of the so-called "empirical rule". This "rule" says that if a random variable X has a symmetric, bell-shaped distribution, then

$$P(|X - \mu_X| \le \sigma_X) \approx 0.68$$

and

$$P(|X - \mu_X| \le 2\sigma_X) \approx 0.95.$$

The aim of this exercise is to examine the limitations of this "rule".

(a) Find a statement of the "rule" in a textbook on elementary statistics.

(b) Suppose that a r.v. has a standard Normal distribution. Verify that the "empirical rule" does indeed work.

(c) Suppose that X has a t-distribution with 3 d.f. Then

$$f_X(t) = \frac{2}{\pi} \cdot \frac{1}{(1+t^2)^2} \quad (t \in \Re).$$

i. Draw the graph of $y = f_X(t)$ and check that it is indeed a bell-shaped curve which is symmetric about $t = 0$.

ii. Verify that $\mu_X = 0$, $\sigma_X = 1$.
iii. By numerical integration, calculate

$$P(|X - \mu_X| \leq \sigma_X)$$

and

$$P(|X - \mu_X| \leq 2\sigma_X).$$

(d) Comment on the utility of the "empirical rule".

(6) Let $X \sim$ Cauchy.

(a) Sketch the graph of the function f_X.
(b) Prove that μ_X does not exist.
(c) Prove that $1/X$ has the same distribution as X.
(d) The property in 6c is rather unusual — but it is not peculiar to the Cauchy distribution. Consider the random variable Y with p.d.f.

$$f_Y(t) = \begin{cases} 1/4 & (|t| \leq 1) \\ 1/(4x^2) & (|t| > 1) \end{cases}$$

Show that $1/Y$ has the same distribution as Y.
(Source: Stoyanov [64, §12.12, p.128])

(7) Problem 6 in Chapter 2.2 dealt with Chebyshev's inequality for discrete random variables. Prove corresponding results for continuous random variables.

(8) We say that the r.v. X has a Pareto distribution with parameter 4 if

$$f_X(t) = \begin{cases} kt^{-4} & (t > 1) \\ 0 & (t \leq 1) \end{cases}$$

where k is some positive constant.

(a) Find the constant k.
(b) Find the d.f. F_X.
(c) Find μ_X, σ_X.
(d) Find $P(|X - \mu_X| \leq 1.5\sigma_X)$.
(e) Use Chebyshev's inequality to estimate $P(|X - \mu_X| \leq 1.5\sigma_X)$ and compare this estimate with the exact value found in the previous part.

(9) Suppose that $\lambda > 0$. If $X \sim \text{Gamma}(\alpha = 1, \beta = \lambda^{-1})$ then we say that X has an exponential distribution with parameter λ.

(a) Write down expressions for $f_X(t)$ and $F_X(t)$.
(b) Sketch the graphs of f_X and F_X.
(c) Let $Y = F_X(X)$. Prove that $Y \sim \text{Uniform}(0, 1)$.
(d) Hence use a random number generator to create a random sample of $n = 5$ observations from an exponential distribution with parameter $\lambda = 2$.

(10) The log normal distribution is important in geostatistics — a branch of statistics dealing with data analysis problems in mineral exploration. (See, for example, [41].) Let $X \sim N(0, 1)$ and define a new r.v. $Y = \exp(X)$. Then we say that Y has a log normal distribution.

(a) Why do we use the name "log-normal"?
(b) Find an expression for $f_Y(t)$.
(c) Sketch the graph of f_Y.

(11) Let $X \sim N(0, 1)$. According to statistical tables, $P(0 \leq X \leq 1) = 0.3413$. Verify this using numerical integration.

(12) **Buffon's needle.** Suppose that a plane is ruled by equidistant parallel lines at distance 1 apart: for example represent the lines by $y = n(n = 0, \pm 1, \pm 2, \pm 3, \ldots)$. A needle of length $L < 1$ is thrown randomly on the plane.

(a) Prove that the probability that the needle will intersect one of the parallel lines is $2L/\pi$.
(b) Hence devise an experiment for estimating π.

[Hint: Let M be the mid-point of the needle. Drop a perpendicular from M to the nearest horizontal line and let the length of this perpendicular be y; then $0 \leq y \leq 1$. Let θ be the acute angle between this perpendicular and the needle; then $0 \leq \theta \leq \pi/2$. We do not have to worry about the end-point possibilities such as $theta = 0$ or $y = 1$ because they have probability 0 of happening.]
(Source: This result is known as "Buffon's" needle problem because it is based on a paper by G. Buffon (1777).)

(13) **Random polynomials.** Suppose that $B \sim \text{Uniform}\,(0, 1)$, $C \sim \text{Uniform}\,(0, 1)$, and that B and C are independent of each other.

Consider the random polynomial

$$p(x) = x^2 + Bx + C.$$

What is the probability that p has real roots?

Chapter 5

Limit theorems

5.1 Notes

Limit theorems are mathematical results which lead to various approximations that can be employed if we have large samples of data. Hence they are very important in applied statistics. At the same time they are often instances of beautiful mathematical results. The purpose of this section is to give you the experience of proving some famous limit theorems yourself. I hope that you enjoy the experience.

5.2 Problems

(1) Many problems in probability require an approximation to $n!$ when n is large. In such situations, one usually uses a formula by James Stirling (1692–1777). Stirling's formula may be stated as follows.

$$\text{As } n \to \infty, \quad n! \sim (2\pi)^{1/2}\, n^{n+1/2} \exp(-n). \tag{5.1}$$

The symbolism in (5.1) means that

$$\lim_{n\to\infty} \frac{n!}{(2\pi)^{1/2}\, n^{n+1/2} \exp(-n)} = 1.$$

Your task is to prove Stirling's formula by following the steps outlined below. (Source: This proof is based on Moran [45, §1.12]. P.A.P. Moran (1917–1988) was a distinguished Australian mathematician. Notes on his life and work can be found in the article by

C.C. Heyde [33].)

(a) Let $S(n) = \log(n!)$ where log denotes the natural logarithm. Show that

$$S(n) = \sum_{p=1}^{n-1} \log(p+1).$$

(b) Let us define

$$\begin{aligned} A(p) &:= \int_p^{p+1} \log t \, dt \\ B(p) &:= (\log(p+1) - \log p)/2 \\ C(p) &:= \int_p^{p+1} \log t \, dt - (\log(p+1) + \log p)/2. \end{aligned}$$

Using these definitions and 1a, show that

$$S(n) = \sum_{p=1}^{n-1}(A(p)+B(p)-C(p)) = (n+1/2)\log n - n + 1 - \sum_{p=1}^{n-1} C(p).$$

(c) From the definition of $C(p)$ in 1b, show that

$$C(p) = (p+1/2)\log\left(\frac{p+1}{p}\right) - 1.$$

(d) Find a source (e.g. a suitable book on calculus) to justify using

$$\log\left(\frac{1+x}{1-x}\right) = 2\sum_{k=1}^{\infty} \frac{x^{2k-1}}{2k-1} \quad (|x| < 1).$$

(e) Using 1c and 1d show that

$$C(p) = \sum_{k=1}^{\infty} \frac{1}{(2k+1)(2p+1)^{2k}}.$$

(f) Show that

$$C(p) < \frac{1}{12p(p+1)} = \frac{1}{12} \cdot \left(\frac{1}{p} - \frac{1}{p+1} \right).$$

(g) Show that

$$C(p) > \frac{1}{3(2p+1)^2} \sum_{k=0}^{\infty} 3^{-k}(2p+1)^{-2k} > \frac{1}{12} \cdot \left(\frac{1}{p+(1/12)} - \frac{1}{p+1+(1/12)} \right).$$

(h) Let

$$D = \sum_{p=1}^{\infty} C(p)$$

$$r_n = \sum_{p=n}^{\infty} C(p).$$

Using 1f and 1g, show that

$$\frac{1}{13} < D < \frac{1}{12} \text{ and } \frac{1}{12n+1} < r_n < \frac{1}{12n}.$$

(i) Using 1b and 1h show that

$$S_n = \left(n + \frac{1}{2}\right) \log n - n + 1 - D + r_n.$$

(j) Using 1a and 1i show that

$$n! = Kn^{(n+1/2)} \exp(-n + r_n)$$

where K is a constant.

(k) Find a source to justify using Wallis' formula:

$$\pi = \lim_{n\to\infty} \frac{2^{4n}(n!)^4}{n(2n!)^2}.$$

(l) Using 1k and 1j show that the constant K in 1j is given by

$$K = \sqrt{2\pi}.$$

(m) Now use 1j and 1l to prove Stirling's formula (5.1).

(n) **An application of Stirling's approximation.** Try to evaluate

$$\binom{200}{100}$$

by using a calculator: most calculators cannot handle this. Then, for comparison, use Stirling's formula to show that

$$\binom{200}{100} \approx \frac{2^{200}}{10\sqrt{\pi}} \approx 9.066 \times 10^{58}.$$

(2) This exercise deals with using the Poisson distribution to approximate binomial distribution probabilities. This approximation is very useful in quality control problems. (See [44].)

(a) Prove that

$$\lim_{n\to\infty}\left(1+\frac{1}{n}\right)^n = e$$

as follows. Let $e_n := \left(1+\frac{1}{n}\right)^n$.

i. Show

$$e_n = \sum_{k=0}^{n} \frac{1}{k!} \prod_{j=1}^{k-1}\left(1-\frac{j}{n}\right). \tag{5.2}$$

ii. Using (5.2), show that $0 < e_1 < e_2 < \ldots$.

iii. Show that

$$\begin{aligned} e_n &< \sum_{k=0}^{n} \frac{1}{k!} \quad \text{by (5.2)} \\ &< 1 + 1 + 2^{-1} + 2^{-2} + \ldots + 2^{n-1} \\ &< 3. \end{aligned}$$

iv. Thus $\lim_{n\to\infty} e_n$ exists and < 3. We call this limit e.

(b) Prove

i. $\lim_{n\to\infty}\left(1-\frac{1}{n^2}\right)^n = 1$;

ii. $\lim_{n\to\infty}\left(1-\frac{1}{n}\right)^n = e^{-1}$;

iii. $\lim_{x\to\infty}\left(1-\frac{1}{x}\right)^x = e^{-1}$;

iv. For $\mu > 0$, $\lim_{n\to\infty}\left(1-\frac{\mu}{n}\right)^n = e^{-\mu}$.

(c) Suppose that X has a binomial distribution with parameters n and p. Furthermore, suppose that we have

$$n \to \infty,$$

and

$$p \to 0$$

but, at all times,

$$np = \mu \quad \text{which is constant.}$$

Prove the following results.

i. For fixed $k = 1, 2, \ldots, n$ we have

$$\begin{aligned} & P(X = k) \\ = \; & \frac{n(n-1)\ldots(n-k+1)}{k!}\left(\frac{\mu}{n}\right)^k\left(1-\frac{\mu}{n}\right)^{n-k} \\ = \; & \frac{n(n-1)\ldots(n-k+1)}{n^k\left(1-\frac{\mu}{n}\right)^k}\left(\frac{\mu^k}{k!}\right)\left(1-\frac{\mu}{n}\right)^n. \end{aligned}$$

ii. For fixed $k = 1, 2, \ldots$ we have

$$\lim_{n\to\infty} P(X = k) = \frac{\exp(-\mu)\mu^k}{k!}.$$

iii. Hence, if $n = 100$ and $p = 0.02$, estimate $P(X \leq 2)$.

(3) **Weak law of large numbers. (WLLN)** Suppose that

$$\{X, X_1, X_2, \ldots\}$$

is a sequence of i.i.d. random variables and

$$E(X) = \mu$$

and

$$\text{Var }(X) = \sigma^2.$$

Let

$$\overline{X} = \frac{1}{n}\sum_{i=1}^{n} X_i\,.$$

(a) Prove that $E(\overline{X}) = \mu$.
(b) Prove that $\text{Var}(\overline{X}) = \sigma^2/n$.
(c) Let $\epsilon > 0$. Using Chebyshev's inequality, prove that

$$P(|\overline{X} - \mu| \geq \epsilon) \leq \frac{\sigma^2}{n\epsilon^2}.$$

(d) Hence prove the WLLN which can be stated as follows.

$$\text{If } \epsilon > 0, \text{ then } \lim_{n\to\infty} P(|\overline{X} - \mu| \geq \epsilon) = 0. \qquad (5.3)$$

(e) In your own words, describe the significance of (5.3).

(4) **Characteristic functions.** Characteristic functions provide us with a useful approach to proving certain limit theorems. This sequence of exercises introduces you to these functions and shows why they are useful. A thorough treatment of the subject can be found in Lukacs [42].
For a r.v. X, we define the characteristic function (c.f.) of X by

$$\phi_X(t) := E(\exp(itX)), \quad (-\infty < t < \infty).$$

(a) Let $X_1, X_2, \ldots, X_n$ be independent r.v. and define

$$S_n = \sum_{i=1}^{n} X_i\,.$$

Prove that

$$\phi_{S_n} = \prod_{i=1}^{n} \phi_{X_i}\,.$$

(b) Let $X, X_1, X_2, \ldots, X_n$ be independent and identically distributed r.v. and define

$$S_n = \sum_{i=1}^{n} X_i .$$

Prove that

$$\phi_{S_n} = (\phi_X)^n.$$

(c) Suppose that the r.v. X, Y are connected by $Y = aX + b$. Prove that

$$\phi_Y(t) = \exp(itb)\phi_X(at).$$

(5) We will often need the characteristic function of a standard Normal random variable. Let $Z \sim N(0,1)$. Prove that

$$\phi_Z(t) = \exp(-t^2/2).$$

(6) **Central limit theorem.** In this exercise, you will prove this famous result by yourself — using some useful tips. A very readable account of the historical development of this theorem can be found in W. Adams [1]. We begin by setting up some machinery — definitions and useful results — which you can use to prove the CLT.

In probability theory there are many different concepts of convergence, so we have to very careful about saying that something converges to something else. Here we introduce the notion of weak convergence.

Definition 5.1 A sequence $\{F_n : n = 1, 2, 3, \ldots\}$ of d.f. converges weakly to a function F if

$$(F : \Re \to \Re \text{ is continuous at } t = t_0) \Rightarrow (\lim_{n\to\infty} F_n(t_0) = F(t_0)).$$

We denote this weak convergence by writing $F_n \Rightarrow F$. (There should be no confusion using $\Rightarrow$ for both "implies" and "converges weakly to".)
The following result is the basis of the "method of characteristic functions". It allows us to prove the weak convergence of d.f. by

proving the convergence of the c.f. (P. Moran [45, p. 252, Theorem 6.2]).

Theorem 5.1 *Let $\{F_n : n = 1, 2, 3, \ldots\}$ be a sequence of d.f. with corresponding c.f. $\{\phi_n : n = 1, 2, 3, \ldots\}$. Then*

$$(\exists d.f.\ F)(F_n \Rightarrow F)$$

if and only if

$$(\exists \phi : \Re \to \Re)(\phi \text{ is continuous at } t = t_0 \Rightarrow \lim_{n\to\infty} \phi_n(t_0) = \phi(t_0)).$$

In this case, ϕ is the c.f. corresponding to the d.f. F.

The next theorem allows us to express the c.f. of certain r.v.s in a useful manner ([54, p. 307, Theorem 9]).

Theorem 5.2 *If Y is a r.v. with $E(Y) = 0$ and $Var(Y) = \sigma_Y^2$ then*

$$\phi_Y(t) = 1 - \sigma_Y^2 t^2/2 + K(t)$$

where

$$\lim_{t\to 0} K(t)/t^2 = 0.$$

Now we can state the central limit theorem.

Theorem 5.3 *If*

- $X, X_1, X_2, \ldots$ *are i.i.d. r.v.s;*
- $\mu = E(X)$;
- $\sigma^2 = Var\ (X)$;
- $S(n) = X_1 + X_2 + \ldots + X_n$;
- $Z(n) = \dfrac{S(n) - E(S(n))}{\sigma_{S(n)}}$;
- F_n *is the d.f. of* $Z(n)$;
- $Z \sim N(0, 1)$;
- $F_Z(t) = \dfrac{1}{\sqrt{2\pi}} \displaystyle\int_{-\infty}^{t} \exp(-u^2/2)\, du$

then

$$(\forall t \in \Re)(\lim_{n\to\infty} F_n(t) = F_Z(t)).$$

We now prove the CLT as follows.

(a) Check that $F_Z(t)$ is continuous for all $t \in \Re$.

(b) Hence, to prove

$$(\forall t \in \Re)(\lim_{n\to\infty} F_n(t) = F_Z(t))$$

(which is our aim), it suffices to prove that $F_n \Rightarrow F_Z$.

(c) By Theorem 5.1 it suffices to prove

$$(\forall t \in \Re)(\lim_{n\to\infty} \phi_{Z_n}(t) = \exp(-t^2/2)).$$

(d) We will need this result. If $\lim_{n\to\infty} \lambda_n = \lambda$ then

$$\lim_{n\to\infty} \left(1 + \frac{\lambda_n}{n}\right)^n = \exp(\lambda).$$

[Hint: Use $\lim_{x\to\infty} \left(1 + \frac{1}{x}\right)^x = e$.]

(e) If $Y_i := X_i - \mu \, (i = 1, 2, 3, \ldots), Y := X - \mu$ then

$$\phi_{Z(n)}(t) = \left(\phi_Y\left(\frac{t}{\sigma\sqrt{n}}\right)\right)^n.$$

(f) Using Theorem 5.2 show that

$$\phi_Y\left(\frac{t}{\sigma\sqrt{n}}\right) = 1 + \frac{1}{n}\left(\frac{-t^2}{2} + H(t,n)\right)$$

where $\lim_{n\to\infty} H(t,n) = 0$.

(g) By 6d, 6e, 6f

$$\lim_{n\to\infty} \phi_{Z(n)}(t) = \exp(-t^2/2).$$

Hence the CLT is proved.

(7) Suppose that we throw a fair die n times and after each throw, we record the score showing as $X_i, (i = 1, 2, 3, \ldots, n)$.

(a) Show that $\mu = E(X_i) = 7/2\,(i = 1, 2, 3, \ldots, n)$.
(b) Show that $\sigma^2 = \text{Var}(X_i) = 35/12\,(i = 1, 2, 3, \ldots, n)$.
(c) Let $\overline{X} = \sum_{i=1}^{n} X_i/n$. Show that $E(\overline{X}) = 7/2$ and $\text{Var}(\overline{X}) = 35/12n$.
(d) Assume that $n = 25$.
   i. Use Chebyshev's inequality to estimate

$$P(|\overline{X} - 3.5| \geq 0.5).$$

   ii. Use the CLT to estimate

$$P(|\overline{X} - 3.5| \geq 0.5).$$

   iii. Compare the two estimates so obtained.
(e) Repeat 7d for $n = 500$.

(8) The owner of a small business is developing a strategic plan for the business and hence, wants to develop some estimates of next year's income. Let $\$X$ denote the weekly income. From last year's records we obtain the estimates:

$$\mu_X = 2000$$

and

$$\sigma_X = 250.$$

(a) Using these estimates and the CLT, estimate the probability that the annual income ($\$A$) will exceed \$ 110,000; that is, find

$$P(A \geq 110,000).$$

(b) Find an interval $[A_1, A_2]$ (with mid-point 104,000) such that

$$P(A_1 \leq A \leq A_2) = 0.90.$$

(Such an interval may be described as a 90% prediction interval for A.)
(c) Discuss any difficulties associated with applying the CLT to this problem.

(9) The following problem is suggested by intelligent style of playing Tattslotto as described in [32, pp. 82–86]. The authors suggest that "a winning combination with a large sum of numbers is likely to go along with a high payoff".
Suppose that we choose 6 numbers

$$\{X_1, X_2, X_3, X_4, X_5, X_6\}$$

at random from the set

$$\{1, 2, 3, \ldots, 44, 45\}.$$

Find a number T such that

$$P(X_1 + X_2 + \ldots + X_6 \geq T) = 0.20.$$

Discuss the limitations of your analysis.

(10) (Source: This problem was motivated by S. Ross [57, p. 423, Problem 14].) A particular component is essential for the operation of an electrical system. In fact, if the component ever fails, it must be replaced immediately. So it is necessary to have replacement components in stock. However, this component is particularly expensive, and hence it is desirable to keep the number of spare components to a minimum. The Quality Engineer recommends that there should be enough components in stock to ensure that the probability that the system stays in operation for $H = 1000$ hours is at least $p = 0.99$.
Let X be the lifetime of a component. Assume that X is a random variable with mean value

$$\mu_X = 50$$

and standard deviation

$$\sigma_X = 15.$$

(a) Find the minimum number (n) of components which should be kept in stock.
(b) What is the effect on n is p is reduced from 0.99 to 0.95?

(c) Establish the following general formula for n in terms of the parameters of the problem H, p, μ_X, and σ_X:

$$n \geq \left(\frac{z_p \sigma_X + \sqrt{z_p^2 \sigma_X^2 + 4\mu_X H}}{2\mu_X} \right)^2$$

where $P(Z \leq z_p) = p$ for $Z \sim N(0,1)$.

(11) Let $X, X_1, X_2, \ldots$ be i.i.d. random variables such that

$$E(X) = 0$$

and

$$\mathrm{Var}(X) = \sigma^2.$$

Let $S(n) = \sum_{i=1}^{n} X_i$. Suppose that

$$\lim_{n\to\infty} P\left(\frac{S(n)}{\sqrt{n}} > 1\right) = \frac{1}{3}.$$

Use CLT to find an estimate of σ^2. (Source: [65].)

(12) This is an application of the method of characteristic functions which is based on Theorem 5.1.

(a) Let

$$\begin{aligned} X(n) &\sim \text{Poisson}(n) \quad (n \in \{1,2,3,\ldots\}); \\ Z(n) &= (X(n) - n)/\sqrt{n}; \\ Z &\sim N(0,1). \end{aligned}$$

Prove that

$$\lim_{n\to\infty} F_{Z(n)}(t) = F_Z(t).$$

(b) Suppose that $X \sim \text{Poisson}(20)$. Use the previous part to estimate $P(15 \leq X \leq 20)$. (Take care in approximating the distribution of a discrete r.v. by the distribution of a continuous r.v.) Calculate this probability exactly and compare your answers.

Chapter 6

Random walks

6.1 Notes

A random walk is a mathematical model. As is so often the case in applied mathematics, a single mathematical model can be used in vastly different contexts — this is the power of mathematics. The random walk has been used to describe fluctuations on financial markets, random motion of particles subject to Brownian motion and the growth of certain populations: an Australian example of the application of random walks to financial markets can be found in Praetz [52]. From a pure mathematical point of view, the subject of random walks is a beautiful theory with many problems which are still unsolved. (See Erdős and Révész [24].) In this chapter, we will explore the simple random walk and, in doing so, we follow Feller [25, Chap. 3]. Another useful source of information is Cox and Miller [19, Chapter 2] More advanced treatments of random walks can be found in Spitzer [62] (— a classic work which has been recently reprinted fortunately) and Révész [55] (— a very attractive work by one of the world's leading scholars on the subject).

We begin with the simple notion of a path in two dimensions.

Definition 6.1 Let $n > 0$ and x be integers. A *path* $(s_0, s_1, \ldots, s_n)$ from $(0, 0)$ to (n, x) is a polygonal line whose vertices are:

$$(0, s_0), (1, s_1) \ldots, (n, s_n)$$

where

$$s_0 = 0, s_n = x, \text{ and } s_k - s_{k-1} = \epsilon_k = \pm 1, \ (k = 1, 2, \ldots, n).$$

Now we define a simple random walk.

Definition 6.2 Let $X, X_1, X_2, \ldots, X_n$ be i.i.d. with

- $P(X = 1) + P(X = -1) = 1$,
- $S_O := 0$,
- $S_n := X_1 + X_2 + \ldots + X_n \quad (n \geq 1)$.

Then $\{S_0, S_1, S_2, \ldots\}$ is a *simple random walk* on the integers.

A particular simple random walk is a symmetric simple random walk.

Definition 6.3 If, in the above definition,

$$P(X = 1) = P(X = -1) = 0.5$$

then we say that $\{S_0, S_1, S_2, \ldots\}$ is a *symmetric simple random walk*, or merely, a symmetric random walk.

Random walks are important simple models to consider in time series analysis. Thus we need the following definition of a general random walk.

Definition 6.4 Let $\{e, e_1, e_2, e_3, \ldots\}$ be i.i.d and $E(e) = 0$, $\text{Var}(e) = \sigma_e^2$. Define a sequence of random variables $\{Z_1, Z_2, \ldots\}$ by

$$\begin{aligned} Z_1 &= e_1 \\ Z_t &= Z_{t-1} + e_t, \quad (t > 1). \end{aligned}$$

Then we say that $\{Z_1, Z_2, \ldots\}$ is a *general random walk* (or sometimes, when there is no confusion, just "random walk").

6.2 Problems

(1) This question explores the definition of a path.

   (a) Prove that there is a path from $(0, 0)$ to $(13, 7)$ but there is no path from $(0, 0)$ to $(13, 8)$.

(b) Prove that there is a path from $(0,0)$ to (n,x) if and only if there exist non-negative integers p,q such that

$$n = p + q \text{ and } x = p - q.$$

[Hint: Let p be the number of $\epsilon_k = +1$ and q be the number of $\epsilon_k = -1$.]

(2) Here, we engage in counting paths.

(a) Show that there are 2^n different paths which start at $(0,0)$ and have length n.

(b) Prove that the number of paths from $(0,0)$ to (n,x) is given by

$$N_{n,x} = \begin{cases} \binom{p+q}{p} & \text{if } n = p+q \text{ and } x = p-q \\ \\ 0 & \text{otherwise.} \end{cases}$$

(c) Prove that the number of paths from (a,b) to (n,x) is $N_{n-a,x-b}$.

(3) **Reflection principle.** This clever idea is quite useful in counting paths.
Let $A = (a,\alpha)$, $B = (b,\beta)$ be integral points with $b > a \geq 0$, $\alpha > 0$, $\beta > 0$, $A' = (a,-\alpha)$. Let S_1 be the set of paths from A to B which touch or cross the horizontal axis, and let S_2 be the set of paths from A' to B.
Prove that $\#S_1 = \#S_2$.
Hint: Establish a 1–1 correspondence between the elements of S_1 and S_2 using reflection — draw a diagram; e.g. the path

$$\{(1,1),(2,2),(3,1),(4,0),(5,1),(6,2)\}$$

corresponds to the path

$$\{(1,-1),(2,-2),(3,-1),(4,0),(5,1),(6,2)\}.$$

(4) Here we apply some of the results found above.

(a) Show that there is a path from (1,1) to (15,5).

(b) How many paths are there from (1,1) to (15,5)?

(c) How many paths are there from (1,1) to (15,5) which touch or cross the horizontal axis?

(d) How many paths are there from (1,1) to (15,5) which do not touch or cross the horizontal axis?

(5) Let $n > 0$, $x > 0$ be integers. Then the number of paths which start at $(0,0)$, do not re-visit the x-axis and end at (n,x) is exactly $(x/n)N_{n,x}$.
Hint: The paths in which we are interested must go through $(1,1)$.

(6) **Ballot theorem.** This is an application of problem 5. Suppose that, in an election, candidate P scores 1500 votes and candidate Q scores 1000 votes. Suppose that, during counting, we define $X_k = 1$ if the k-th vote is for P and $X_k = -1$ if the k-th vote is for Q.

(a) Show that $S_n := X_1 + X_2 + \ldots + X_n$ is a simple random walk.
(b) What is the significance of the event $S_n > 0$?
(c) Show that the probability that, throughout the counting of votes, P is leading Q equals 0.2.
(d) Suppose that, in the above problem, we replaced 1500 by p and 1000 by q. Show that the probability that, throughout the counting of votes, P is leading Q equals $(p-q)/(p+q)$.

(7) Let $\{S_0, S_1, S_2, \ldots\}$ be a symmetric random walk. Prove that

$$P(S_n = r) = \binom{n}{(n+r)/2} 2^{-n}. \tag{6.1}$$

where $\binom{n}{(n+r)/2}$ is taken to be zero unless $(n+r)/2 \in \{0, 1, \ldots, n\}$. Establish an analogous formula for a non-symmetric, simple random walk. (Hint: Use the binomial distribution.)

(8) Here is a limit theorem in random walks.

(a) Prove that, for a symmetric random walk,

$$P(S_{2\nu} = 0) \sim (\sqrt{\pi\nu})^{-1} \text{ as } \nu \to \infty.$$

[Hints: The symbol $\sim$ was defined in equation (5.1). Use equation (6.1) and Stirling's formula.]

(b) Investigate the relative error of this approximation for modest values of ν; i.e. calculate

$$\left(\frac{P(S_{2\nu} = 0) - (\sqrt{\pi\nu})^{-1}}{P(S_{2\nu} = 0)}\right) \times 100\%$$

for modest values of ν.

(9) The road to ruin. (Source: [27].)

(a) A fair coin is tossed repeatedly. Show that a head is bound to turn up eventually. (You need to turn this last statement into a formal mathematical statement.)
(b) A fair coin is tossed repeatedly. Show that a sequence of 3 consecutive heads is bound to turn up eventually.
(c) A fair coin is tossed repeatedly. Show that any given finite sequence of heads and tails is bound to turn up eventually.
(d) A symmetric random walk takes place on the integers starting at 5. Show that the walk is bound to reach 0 or 10 eventually.
(e) Two gamblers A and B play the following game. A fair coin is tossed repeatedly;

- whenever it comes down *Heads* then A wins \$1 from B;
- whenever it comes down *Tails* then B wins \$1 from A.

Assume that each player starts with \$5. The game stops when one player has no money left. Show that the game is bound to stop eventually.

(10) Let $\{Z_1, Z_2, \ldots\}$ be a general random walk. Prove the following results.

(a) $Z_t = e_1 + e_2 + \ldots e_t$ for $t \geq 1$.
(b) $E(Z_t) = 0$ for $t \geq 1$.
(c) Var $(Z_t) = t\sigma_e^2$ for $t \geq 1$.
(d) Cov $(Z_t, Z_s) := E(Z_t Z_s) - E(Z_t)E(Z_s) = t\sigma_e^2$ for $1 \leq t \leq s$.
(e) Corr $(Z_t, Z_s) = \sqrt{t/s}$ for $1 \leq t \leq s$.
(f) $\lim_{t\to\infty}$ Corr $(Z_t, Z_{t+1}) = 1$.

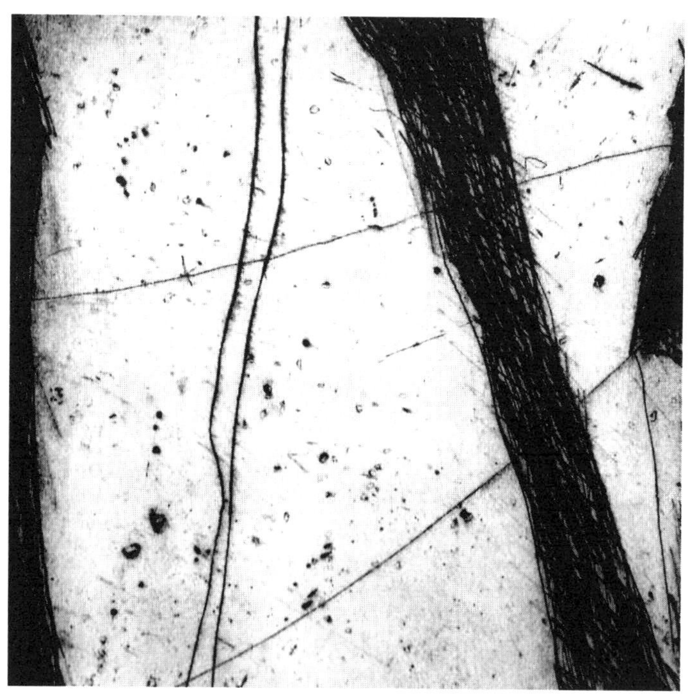

PART 2

Solutions

Chapter 1

Sets, measure and probability

(1) It is not good teaching practice to start off with an advanced problem. However, this problem is particularly intriguing because, although it deals with the most elementary notions of set theory, it throws you immediately into a fascinating topic in mathematics (namely the study of Dedekind numbers) in which there is recent research activity. It leads you also to the wonderful world of sequences. According to N.J.A. Sloane's site, the next number in the sequence is

$$2,414,682,040,996$$

— as you probably guessed!

(2) Let A, B, C, D be subsets of Ω.

(a) $((A\cap B)\cup(C\cap D))' = (A\cap B)'\cap(C\cap D)' = (A'\cup B')\cap(C'\cup D')$.

(b) Let

$$x \in (A\cup B)\cap(A\cup B')\cap(A'\cup B)\cap(A'\cup B').$$

Therefore, x is an element of the intersection of 4 sets each of which is the union of 2 sets. This leads to 16 possibilities — each of which is impossible (e.g. $x \in A\cap A\cap A'\cap A' = \emptyset$ or $x \in B\cap A\cap B\cap B' = \emptyset$). A tree diagram depicting all 16 contingencies may be helpful.

(c) Using the definition of $\triangle$ we obtain

$$\begin{aligned} A\triangle\Omega &= (A\cap\Omega')\cup(A'\cap\Omega) \\ &= (A\cap\emptyset)\cup(A'\cap\Omega) \end{aligned}$$

$$\begin{aligned} &= \emptyset \cup A' \\ &= A'. \end{aligned}$$

(d) Using the definition of $\triangle$ we obtain

$$\begin{aligned} A \cap (A \triangle B) &= A \cap ((A \cap B') \cup (A' \cap B)) \\ &= (A \cap A \cap B') \cup (A \cap A' \cap B) \\ &= (A \cap B') \cup (\emptyset \cap B) \\ &= (A \setminus B) \cup \emptyset \\ &= (A \setminus B). \end{aligned}$$

(e) The sets $A \triangle B$ and $A \cap B$ are disjoint. So

$$\begin{aligned} (A \triangle B) \triangle (A \cap B) &= (A \triangle B) \cup (A \cap B) \\ &= (A \cap B') \cup (A' \cap B) \cup (A \cap B) \\ &= (A \cap B') \cup (B \cap (A' \cup A)) \\ &= (A \cap B') \cup B \\ &= (A \cup B) \cap (B' \cup B) \\ &= A \cup B. \end{aligned}$$

(f) To solve such a problem, reduce one side to an expression which is not changed by interchanging A, B, C.

$$\begin{aligned} & A \triangle (B \triangle C) \\ = \;& (A \cap (B \triangle C)') \cup (A' \cap (B \triangle C)) \\ = \;& (A \cap ((B \cap C') \cup (B' \cap C))') \\ & \quad \cup (A' \cap ((B' \cap C) \cup (B \cap C'))) \\ = \;& (A \cap ((B \cap C')' \cap (B' \cap C)')) \\ & \quad \cup ((A' \cap B' \cap C) \cup (A' \cap B \cap C')) \\ = \;& (A \cap ((B' \cup C) \cap (B \cup C'))) \\ & \quad \cup ((A' \cap B' \cap C) \cup (A' \cap B \cap C')) \\ = \;& (A \cap (B' \cup C) \cap (B \cup C')) \\ & \quad \cup ((A' \cap B' \cap C) \cup (A' \cap B \cap C')) \\ = \;& (((A \cap B') \cup (A \cap C)) \cap (B \cup C')) \\ & \quad \cup ((A' \cap B' \cap C) \cup (A' \cap B \cap C')) \end{aligned}$$

$$
\begin{aligned}
&= (((A \cap B') \cap (B \cup C')) \cup ((A \cap C) \cap (B \cup C'))) \\
&\quad \cup((A' \cap B' \cap C) \cup (A' \cap B \cap C')) \\
&= (A \cap B' \cap B) \cup (A \cap B' \cap C') \\
&\quad \cup(A \cap B \cap C)) \cup (A \cap C \cap C') \\
&\quad \cup(A' \cap B' \cap C) \cup (A' \cap B \cap C') \\
&= (A \cap B \cap C) \cup (A \cap B' \cap C') \\
&\quad \cup(A' \cap B \cap C') \cup (A' \cap B' \cap C).
\end{aligned}
$$

Interchanging the symbols A, B, C has no effect on the last line; so interchanging the symbols A, B, C has no effect on the first line. Thus

$$A \triangle (B \triangle C) = C \triangle (A \triangle B) = (A \triangle B) \triangle C.$$

(g) There are two approaches to this problem.
A straight forward approach is as follows.

$$
\begin{aligned}
&(A \cap B') \triangle (B \cap A') \\
&= ((A \cap B') \cap (B \cap A')') \cup ((A \cap B')' \cap (B \cap A')) \\
&= ((A \cap B') \cap (B' \cup A)) \cup ((A' \cup B) \cap (B \cap A')) \\
&= (A \cap B' \cap B') \cup (A \cap B' \cap A) \\
&\quad \cup(A' \cap B \cap A') \cup (B \cap B \cap A') \\
&= (A \cap B') \cup (A \cap B') \cup (A' \cap B) \cup (A' \cap B) \\
&= (A \cap B') \cup (A' \cap B) \\
&= A \triangle B.
\end{aligned}
$$

A slicker approach is to observe that $A \cap B'$ and $B \cap A'$ are disjoint. So

$$(A \cap B') \triangle (B \cap A') = (A \cap B') \cup (B \cap A') = A \triangle B.$$

(h) Let $x \in A \triangle C$. There are four cases to consider.

(1) $x \in A, x \notin C, x \in B$
(2) $x \in A, x \notin C, x \notin B$
(3) $x \notin A, x \in C, x \in B$
(4) $x \notin A, x \notin C, x \notin B$

Assume case 1 holds. Then, assuming $A\triangle B = C\triangle D$, we have

$$\begin{aligned} & x \in A, x \notin C, x \in B \\ \Rightarrow \quad & x \in A, x \in B, x \notin C, x \notin A\triangle B \\ \Rightarrow \quad & x \in A, x \in B, x \notin C, x \notin C\triangle D \\ \Rightarrow \quad & x \in B, x \notin C, x \notin C\triangle D \\ \Rightarrow \quad & x \in B, x \notin C \cup (C\triangle D) \\ \Rightarrow \quad & x \in B, x \notin D \\ \Rightarrow \quad & x \in B\triangle D. \end{aligned}$$

Similar arguments work for the other cases.
Thus $A\triangle C \subset B\triangle D$.
Similarly $B\triangle D \subset A\triangle C$.
Thus $A\triangle C = B\triangle D$.

(i) This problem demonstrates that $\cap$ is distributive over $\triangle$.

$$\begin{aligned} & (A\cap B)\triangle(A\cap C) \\ = \quad & ((A\cap B)' \cap (A\cap C)) \cup ((A\cap B)\cap (A\cap C)') \\ = \quad & ((A'\cup B')\cap (A\cap C)) \cup ((A\cap B)\cap (A'\cup C')) \\ = \quad & ((A\cap C)\cap (A'\cup B')) \cup ((A\cap B)\cap (A'\cup C')) \\ = \quad & ((A\cap C\cap A')\cup (A\cap C\cap B')) \\ & \quad \cup((A\cap B\cap A')\cup (A\cap B\cap C')) \\ = \quad & (A\cap B'\cap C)\cup (A\cap B\cap C')) \\ = \quad & A\cap ((B'\cap C)\cup (B\cap C')) \\ = \quad & A\cap (B\triangle C). \end{aligned}$$

(j) The associativity of $\triangle$, as proved above, allows us to dispense with parentheses on the right hand side. Also note that, for any set X, $X\triangle X = \emptyset$ and $X\triangle\emptyset = X$. Therefore,

$$(A\triangle C)\triangle(C\triangle B) = A\triangle C\triangle C\triangle B = A\triangle\emptyset\triangle B = A\triangle B.$$

(3) Arguments using indicator functions are nice alternatives to set theoretic arguments.

(a) This result allows us to prove identities between sets by checking that the indicator functions are equal.

$$\begin{aligned} & A = B \\ \Leftrightarrow \quad & (\forall x \in \Omega)(x \in A \Leftrightarrow x \in B) \\ \Leftrightarrow \quad & (\forall x \in \Omega)(I_A(x) = I_B(x)) \\ \Leftrightarrow \quad & I_A = I_B. \end{aligned}$$

One can also say that $(A = B) \Leftrightarrow (I_A \equiv I_B (\text{mod } 2))$. This will be useful below.

(b) Here we demonstrate the correspondence between set operations and algebraic operations on indicator functions.

i. From the definition of an indicator function,

$$\begin{aligned} I_\Omega(x) &= \begin{cases} 1 & \text{if } x \in \Omega \\ 0 & \text{if } x \in \Omega' = \emptyset. \end{cases} \\ &= 1 \text{ if } x \in \Omega. \end{aligned}$$

So, $I_\Omega = 1$. Similarly $I_\emptyset = 0$.

ii. Here we see that $\cap$ corresponds to multiplication.

$$\begin{aligned} & x \in A \cap B \\ \Leftrightarrow \quad & (x \in A) \wedge (x \in B) \\ \Leftrightarrow \quad & I_A(x) I_B(x) = 1. \end{aligned}$$

Thus, $I_{A \cap B} = I_A I_B$

iii. There are 4 cases to consider, namely $x \in A \backslash B$, $x \in B \backslash A$, $x \in A \cap B$, $x \in (A \cup B)'$. We show that, in each case $I_{A \cup B} = I_A + I_B - I_A I_B$. We illustrate with just one case. Suppose that $x \in B \backslash A$. Then $I_{A \cup B}(x) = 1$. On the other hand,

$$I_A(x) + I_B(x) - I_A(x) I_B(x) = 0 + 1 - (0)(1) = 1.$$

Thus $I_{A \cup B}(x) = I_A(x) + I_B(x) - I_A(x) I_B(x)$.

iv. Here we see that complementation corresponds to "1−".

$$I_{A'}(x) = \begin{cases} 1 & \text{if } x \in A' \\ 0 & \text{if } x \in (A')' \end{cases}$$

$$= \begin{cases} 0 & \text{if } x \in A \\ 1 & \text{if } x \in A' \end{cases}$$
$$= 1 - I_A(x).$$

v. There are 4 cases to consider, namely $x \in A \setminus B$, $x \in B \setminus A$, $x \in A \cap B$, $x \in (A \cup B)'$. We show that, in each case $I_{A \triangle B}(x) \equiv I_A(x) + I_B(x) \pmod 2$. We illustrate with just one case.
Suppose that $x \in A \cap B$. Then $I_{A \triangle B}(x) = 0$. Also $I_A(x) + I_B(x) = 2 \equiv 0 \pmod 2$. So $I_{A \triangle B}(x) \equiv I_A(x) + I_B(x) \pmod 2$.

vi. $I_{A \setminus B} = I_{A \cap B'} = I_A I_{B'} = I_A(1 - I_B)$.

(c) This is where indicator functions flex their muscles. To show that two sets are equal, it suffices to show that their indicator functions are equal. We illustrate this with just a few examples from Problem 1. It will be more convenient to write $I(A)$ rather than I_A. Remember that $I(A)^2 = I(A)$ and that $I(A)I(A') = I(A \cap A') = 0$.

1b This problem is much easier to solve with indicator functions.

$$\begin{aligned} & I((A \cup B) \cap (A \cup B') \cap (A' \cup B) \cap (A' \cup B')) \\ = \; & I(A \cup B) I(A \cup B') I(A' \cup B) I(A' \cup B') \\ = \; & (I(A) + I(B) - I(A)I(B)) \cdot \\ & \quad (I(A) + I(B') - I(A)I(B')) \cdot \\ & \quad (I(A') + I(B) - I(A')I(B)) \cdot \\ & \quad (I(A') + I(B') - I(A')I(B')) \\ = \; & \vdots \\ = \; & I(A)I(A') = I(\emptyset). \end{aligned}$$

1f This makes associativity of $\triangle$ easy to prove!

$$\begin{aligned} I(A \triangle (B \triangle C)) & \equiv I(A) + I(B \triangle C) \pmod 2 \\ & \equiv I(A) + (I(B) + I(C)) \pmod 2 \\ & \equiv (I(A) + I(B)) + I(C) \pmod 2 \\ & \equiv I(A \triangle B) + I(C) \pmod 2 \end{aligned}$$

$$\equiv \quad I((A \triangle B) \triangle C) \pmod 2.$$

So $I(A \triangle (B \triangle C)) = I((A \triangle B) \triangle C)$.

1h We must show that

$$I(A \triangle B) = I(C \triangle D) \Rightarrow I(A \triangle C) = I(B \triangle D).$$

$$\begin{aligned} I(A \triangle B) &= I(C \triangle D) \\ I(A) + I(B) &\equiv I(C) + I(D) \pmod 2 \\ I(A) - I(C) &\equiv -I(B) + I(D) \pmod 2 \\ I(A) + I(C) &\equiv I(B) + I(D) \pmod 2 \\ I(A \triangle C) &= I(B \triangle D). \end{aligned}$$

(4) Let Ω be a set and suppose that $\mathcal{R}$ is a non-empty set of subsets of Ω. We say that $\mathcal{R}$ is a **ring** of subsets of Ω if

$$(A \in \mathcal{R} \text{ and } B \in \mathcal{R}) \Rightarrow (A \cup B \in \mathcal{R} \text{ and } A \setminus B \in \mathcal{R}).$$

(a) Since $\mathcal{R}$ is non-empty, let $A \in \mathcal{R}$. Then $A \setminus A = \emptyset \in \mathcal{R}$.

(b) $\Omega = \{1, 2, 3, 4\}$; $A = \{1, 2\}$; $\mathcal{R} = \{\emptyset, A\}$.

(c) Assume that $\mathcal{R}$ is a ring; let $A \in \mathcal{R}, B \in \mathcal{R}$. Then $A \cup B \in \mathcal{R}$, $A \setminus B \in \mathcal{R}$, $B \setminus A \in \mathcal{R}$, $(A \setminus B) \cup (B \setminus A) = A \triangle B \in \mathcal{R}$; $(A \cup B) \setminus (A \triangle B) = A \cap B \in \mathcal{R}$.
Conversely, assume

$$(A \in \mathcal{R} \text{ and } B \in \mathcal{R}) \Rightarrow (A \cap B \in \mathcal{R} \text{ and } A \triangle B \in \mathcal{R}).$$

Let $A \in \mathcal{R}$ and $B \in \mathcal{R}$. Then $A \triangle B \in \mathcal{R}$; $A \cap B \in \mathcal{R}$; $A \cap (A \triangle B) = A \setminus B \in \mathcal{R}$; $A \cup B = (A \triangle B) \triangle (A \cap B) \in \mathcal{R}$.
So $\mathcal{R}$ is a ring of sets.

(d) A counterexample will suffice. Let $\Omega = \{1, 2, 3, 4\}$; $A = \{1, 2\}$; $A' = \{3, 4\}$; $S = \{\emptyset, A, A'\}$. Then S is not a ring because $\Omega = A \cup A' \notin S$.

(e) The following argument is commonly used in mathematics where we have to prove that a system of sets is closed under certain operations. Let $\mathcal{R}_1, \mathcal{R}_2$ be two rings of subsets of Ω. Let $\mathcal{R} = \mathcal{R}_1 \cap \mathcal{R}_2$. Then

$$\begin{aligned} & A \in \mathcal{R}, B \in \mathcal{R} \\ \Rightarrow \quad & (A \in \mathcal{R}_1 \text{ and } B \in \mathcal{R}_1) \text{ and } (A \in \mathcal{R}_2 \text{ and } B \in \mathcal{R}_2) \end{aligned}$$

$$\begin{aligned} \Rightarrow \quad & (A \cup B \in \mathcal{R}_1 \text{ and } A \setminus B \in \mathcal{R}_1) \\ & \quad \text{and } (A \cup B \in \mathcal{R}_2 \text{ and } A \setminus B \in \mathcal{R}_2) \\ \Rightarrow \quad & A \cup B \in \mathcal{R} \text{ and } A \setminus B \in \mathcal{R}. \end{aligned}$$

Thus $\mathcal{R}$ is a ring.

(5) For details on Kolmogorov, see [2], [48]; his most famous work is [40].

(6) Here we prove from the axioms some results which may seem obvious. However, it is important to demonstrate that our axioms lead to these results.

(a) Since $\mathcal{A}$ is a σ-algebra, $\Omega \in \mathcal{A}$; hence $\Omega' = \emptyset \in \mathcal{A}$; so $P(\emptyset)$ is well defined.
(b) $1 = P(\Omega) = P(\Omega \cup \emptyset) = P(\Omega) + P(\emptyset) = 1 + P(\emptyset)$; so $P(\emptyset) = 0$.
(c) $1 = P(\Omega) = P(A \cup A') = P(A) + P(A')$; so $P(A') = 1 - P(A)$.
(d) Observe that $A \cap B \in \mathcal{A}$ because $A \cap B = (A' \cup B')'$.
(e) We can express $A \cup B$ as a disjoint union:

$$A \cup B = (A \cap B') \cup (A \cap B) \cup (A' \cap B),$$

and therefore,

$$\begin{aligned} & P(A \cup B) \\ = \quad & P(A \cap B') + P(A \cap B) + P(A' \cap B) \\ = \quad & P(A \cap B') + P(A \cap B) + P(A' \cap B) \\ & \quad + P(A \cap B) - P(A \cap B) \\ = \quad & P(A) + P(B) - P(A \cap B). \end{aligned}$$

(7) Here we use the above expression for $P(A \cup B)$ repeatedly.

(a) We write the union of 3 sets as union of 2 sets and then employ the result above for the probability of the union of 2 sets.

$$\begin{aligned} & P(A \cup B \cup C) \\ = \quad & P((A \cup B) \cup C) \\ = \quad & P(A \cup B) + P(C) - P((A \cup B) \cap C) \\ = \quad & P(A \cup B) + P(C) - P((A \cap C) \cup (B \cap C)) \\ = \quad & P(A) + P(B) - P(A \cap B) + P(C) \end{aligned}$$

$$
\begin{aligned}
&-(P(A\cap C)+P(B\cap C)-P(A\cap C\cap B\cap C))\\
= \quad & P(A)+P(B)+P(C)\\
&-P(A\cap B)-P(A\cap C)-P(B\cap C)\\
&+P(A\cap B\cap C).
\end{aligned}
$$

(b) Look at the pattern in the previous question and guess that:

$$
\begin{aligned}
& P(A\cup B\cup C\cup D)\\
= \quad & P(A)+P(B)+P(C)+P(D)\\
&-P(A\cap B)-P(A\cap C)-P(A\cap D)\\
&-P(B\cap C)-P(B\cap D)-P(C\cap D)\\
&+P(B\cap C\cap D)+P(A\cap C\cap D)\\
&+P(A\cap B\cap D)+P(A\cap B\cap C)\\
&-P(A\cap B\cap C\cap D).
\end{aligned}
$$

While the essence of mathematical argument is proof, there is also a role for guessing the right answer to be proved!

(c) The required result for n sets is

$$
\begin{aligned}
& P(A_1\cup A_2\cap A_3\cup\ldots\cup A_n)\\
= \quad & \sum_{k=1}^{n}(-1)^{k-1}\sum_{1\le j_1<j_2<\ldots<j_k\le n} P(A_{j_1}\cup A_{j_2}\cup\ldots\cup A_{j_k}).
\end{aligned}
$$

(8) Boole's inequality gives an upper bound for the probability of the union of a number of events.

(a) $P(A\cup B)=P(A)+P(B)-P(A\cap B)\le P(A)+P(B)$.

(b) Using the previous result for 2 sets we obtain:

$$
\begin{aligned}
P(A\cup B\cup C) &= P((A\cup B)\cup C)\\
&\le P(A\cup B)+P(C)\\
&\le P(A)+P(B)+P(C).
\end{aligned}
$$

(c) Take the hint from the previous answer and use induction to prove

$$P(A_1 \cup A_2 \cup \ldots A_n) \leq \sum_{i=1}^{n} P(A_i).$$

(9) This is another problem dealing with inequalities in probability.

(a) Since

$$P(A \cup B) = P(A) + P(B) - P(A \cap B)$$

we have

$$\begin{aligned} P(A \cap B) &= P(A) + P(B) - P(A \cup B) \\ &\geq P(A) + P(B) - 1. \end{aligned}$$

(b) Using the previous part we obtain

$$\begin{aligned} P(A \cap B \cap C) &= P((A \cap B) \cap C) \\ &\geq P((A \cap B) + P(C) - 1 \\ &\geq P(A) + P(B) - 1 + P(C) - 1 \\ &= P(A) + P(B) + P(C) - 2. \end{aligned}$$

(c) Taking the hint from the previous answer, we use induction to obtain

$$P(\cap_{i=1}^{n} A_i) \geq \sum_{i=1}^{n} P(A_i) - (n-1).$$

(10) More inequalities in probability.

(a) Let A, B be events.

$$\begin{aligned} P(A \cup B) &= P(A) + P(B) - P(A \cap B) \\ &= 1 - P(A') + P(B) - P(A \cap B) \\ &= 1 - P(A') + P(B \setminus A) \\ &\geq 1 - P(A') \\ &\geq 1 - P(A') - P(B'). \end{aligned}$$

(b) Using the result of the previous part, we obtain

$$\begin{aligned} P(A_1 \cup A_2 \cup A_3) &= P((A_1 \cup A_2) \cup A_3) \\ &\geq 1 - P((A_1 \cup A_2)') - P(A_3') \\ &= P(A_1 \cup A_2) - P(A_3') \\ &\geq 1 - P(A_1') - P(A_2') - P(A_3'). \end{aligned}$$

(c) By induction we prove

$$P(A_1 \cup A_2 \cup \ldots \cup A_n) \geq 1 - \sum_{i=1}^{n} P(A_i') \quad (n = 1, 2, \ldots).$$

(11) We use the following results:

$$A \triangle C = (A \triangle B) \triangle (B \triangle C)$$

and

$$P(A \triangle B) = P(A \setminus B) + P(B \setminus A) \leq P(A) + P(B).$$

Then

$$P(A \triangle C) = P((A \triangle B) \triangle (B \triangle C)) \leq P(A \triangle B) + P(B \triangle C).$$

Alternatively, one may observe that

$$A \triangle C \subset (A \triangle B) \cup (B \triangle C)$$

and hence

$$P(A \triangle C) \leq P(A \triangle B) + P(B \triangle C)$$

and the required triangle inequality is proved.

(12) We represent the events A, B, C and various probabilities $p, q, \ldots, w$ by the following Venn diagram.

	A		A'	
	C	C'	C	C'
B	p	q	r	s
B'	t	u	v	w

So $P(A \cap B \cap C) = p$, $P(A \cap B \cap C') = q$ etc. There are four cases to consider.

Case 1. Suppose that $P(A \cup C) = 0$.

Then $d(A, C) = 0$ and obviously

$$d(A, C) \leq d(A, B) + d(B, C).$$

Case 2. Suppose that $P(A \cup C) \neq 0$ and $P(A \cup B) = 0$.

Then $p = q = t = u = r = s = 0$, $v \neq 0$. In this case,

$$d(A, C) = v/v = 1, \quad d(A, B) = 0, \quad d(B, C) = v/v = 1.$$

So

$$d(A, C) \leq d(A, B) + d(B, C).$$

Case 3. Suppose that $P(A \cup C) \neq 0$ and $P(B \cup C) = 0$. This is similar to Case 2.

Case 4. Suppose that

$$\begin{aligned} P(A \cup C) &\neq 0, \\ P(A \cup B) &\neq 0 \quad \text{and} \\ P(B \cup C) &\neq 0. \end{aligned}$$

Then the following inequalities are equivalent.

$$\frac{P(A \triangle C)}{P(A \cup C)} \leq \frac{P(A \triangle B)}{P(A \cup B)} + \frac{P(B \triangle C)}{P(B \cup C)};$$

$$\begin{aligned} &\frac{q + u + r + v}{p + q + t + u + r + v} \\ \leq\ &\frac{t + u + r + s}{p + q + t + u + r + s} + \frac{q + s + t + v}{p + q + r + s + t + v}; \end{aligned}$$

$$\begin{aligned} &\frac{q + u + r + v}{x - s} \\ \leq\ &\frac{r + s + t + u}{x - v} + \frac{q + s + t + v}{x - u} \\ &\text{where } x = p + q + r + s + t + u + v; \end{aligned}$$

$$(r + u) + (q + v)$$

$$\leq \quad (r+u)\left(\frac{x-s}{x-v}\right)+(q+v)\left(\frac{x-s}{x-u}\right) +(s+t)\left(\frac{x-s}{x-v}+\frac{x-s}{x-u}\right);$$

$$(r+u)\left(\frac{v-s}{x-v}\right)+(q+v)\left(\frac{u-s}{x-u}\right) +(s+t)\left(\frac{x-s}{x-v}+\frac{x-s}{x-u}\right) \quad \geq \quad 0;$$

$$\frac{v(r+u)}{x-v}+\frac{u(q+v)}{x-u} +s\left(\frac{x-s}{x-v}+\frac{x-s}{x-u}-\frac{r+u}{x-v}-\frac{q+v}{x-u}\right) + \quad t\left(\frac{x-s}{x-v}+\frac{x-s}{x-u}\right) \quad \geq \quad 0;$$

$$\frac{v(r+u)}{x-v}+\frac{u(q+v)}{x-u} +s\left(\frac{x-(r+s+u)}{x-v}+\frac{x-(q+s+v)}{x-u}\right) + \quad t\left(\frac{x-s}{x-v}+\frac{x-s}{x-u}\right) \quad \geq \quad 0.$$

This last inequality is true by virtue of the fact that the right hand side is a sum of positive terms.

Thus, the triangle inequality is proved for all cases.

(13) The set $\mathcal{A}$ is closed under the operations of union and intersection because

$$\cup_{\lambda\in\Lambda}(x_\lambda,\infty)=(\inf x_{\lambda\in\Lambda},\infty)\in\mathcal{A}$$

and

$$\cap_{\lambda\in\Lambda}(x_\lambda,\infty)=(\sup x_{\lambda\in\Lambda},\infty)\in\mathcal{A}.$$

However $\mathcal{A}$ is not closed under the operation of complementation because

$$\Omega \setminus (x, \infty) = (-\infty, x] \notin \mathcal{A}.$$

(14) Note that:

$$\begin{aligned} & (\exists X)((A \cap X) \cup (B \cap X') = \emptyset) \\ \Leftrightarrow\ & (\exists X)((A \cap X = \emptyset) \wedge (B \cap X' = \emptyset)) \\ \Leftrightarrow\ & (\exists X)((A \subset X') \wedge (X' \subset B')) \\ \Leftrightarrow\ & (\exists X)((X \subset A') \wedge (B \subset X)) \\ \Leftrightarrow\ & (\exists X)(B \subset X \subset A'). \end{aligned}$$

Thus the necessary and sufficient condition required is that $B \subset A'$.

(15) $A \cap B = \emptyset$; $A \subset B'$; $P(A) \leq P(B')$.

(16) This problem deals with limiting operations applied to sets.

(a) Let $B_n = A_n \setminus (\cup_{k=1}^{n-1} A_k)$. Then it follows that:

$$B_m \cap B_n = \emptyset \quad (\text{if } m \neq n);$$

$$\cup_{k=1}^{n} B_k = \cup_{k=1}^{n} A_k = A_n;$$

$$\cup_{k=1}^{\infty} B_k = \cup_{k=1}^{\infty} A_k = A.$$

Thus,

$$\begin{aligned} P(A) &= P(\cup_{k=1}^{\infty} A_k) \\ &= P(\cup_{k=1}^{\infty} B_k) \\ &= \sum_{k=1}^{\infty} P(B_k) \\ &= \lim_{n \to \infty} \sum_{k=1}^{n} P(B_k) \\ &= \lim_{n \to \infty} P(\cup_{k=1}^{n} B_k) \\ &= \lim_{n \to \infty} P(A_n). \end{aligned}$$

(b) If $A_1 \supset A_2 \supset \ldots$ then $A'_1 \subset A'_2 \subset \ldots$. We will apply the previous part of this question to the sequence $\{A'_n : n = 1, 2, 3, \ldots\}$. So:

$$\begin{aligned} P(A') &= P\left((\cap_{n=1}^{\infty} A_n)'\right) \\ &= P\left(\cup_{n=1}^{\infty} (A'_n)\right) \\ &= \lim_{n\to\infty} P(A'_n) \\ 1 - P(A) &= 1 - \lim_{n\to\infty} P(A_n) \\ P(A) &= \lim_{n\to\infty} P(A_n). \end{aligned}$$

(c) Here the notions of lim sup and lim inf are applied to a sequence of sets.

i. Observe that

$$\begin{aligned} & x \in \limsup A_n \\ \Leftrightarrow\ & x \in \cap_{j=1}^{\infty} \cup_{i=j}^{\infty} A_i \\ \Leftrightarrow\ & (\forall j)(x \in \cup_{i=j}^{\infty} A_i) \\ \Leftrightarrow\ & (\forall j)(\exists i)((i \geq j) \wedge (x \in A_i)) \\ \Leftrightarrow\ & x \in A_k \text{ for infinitely many values of } k. \end{aligned}$$

ii. Observe that

$$\begin{aligned} & x \in \liminf A_n \\ \Leftrightarrow\ & x \in \cup_{j=1}^{\infty} \cap_{i=j}^{\infty} A_i \\ \Leftrightarrow\ & (\exists j)(x \in \cap_{i=j}^{\infty} A_i) \\ \Leftrightarrow\ & (\exists j)(\forall i)((i \geq j) \Rightarrow (x \in A_i)) \\ \Leftrightarrow\ & x \in A_k \text{ for all but finitely many values of } k. \end{aligned}$$

iii. The two previous parts imply that $\liminf A_n \subset \limsup A_n$.

iv. We solve this using elementary symbolic logic. Like abstract art, this style appeals to some and not to others. An excellent introduction to symbolic logic is Copi [17].

$$\begin{aligned} & x \in \liminf(A'_n) \\ \Leftrightarrow\ & (\exists j)(\forall i)(i \geq j \Rightarrow x \in A'_i) \\ \Leftrightarrow\ & (\exists j)(\forall i)(i \geq j \Rightarrow x \notin A_i) \end{aligned}$$

$$\begin{aligned}
&\Leftrightarrow \quad (\exists j)(\forall i)(\neg((i \geq j) \wedge (x \in A_i))) \\
&\Leftrightarrow \quad \neg(\forall j)(\exists i)((i \geq j) \wedge (x \in A_i)) \\
&\Leftrightarrow \quad \neg(x \in \limsup A_n) \\
&\Leftrightarrow \quad x \in (\limsup A_n)'
\end{aligned}$$

v. Let $x \in \limsup(A_n \triangle A_{n+1})$. Then x is an element of infinitely many of the events $A_n \triangle A_{n+1}$. Suppose that k is one of the infinitely many indices such that $x \in A_k \triangle A_{k+1}$. Then either (a) $x \in A_k$ and $x \notin A_{k+1}$ or (b) $x \notin A_k$ and $x \in A_{k+1}$. Thus

- x is an element of infinitely many of the events $\{A_n : n = 1, 2, 3, \ldots\}$, and,
- x is an element of infinitely many of the events $\{A'_n : n = 1, 2, 3, \ldots\}$.

So

$$x \in \limsup A_n \text{ but } x \notin \liminf A_n;$$

that is,

$$x \in (\limsup A_n) \setminus (\liminf A_n).$$

Thus

$$\limsup(A_n \triangle A_{n+1}) \subset (\limsup A_n) \setminus (\liminf A_n).$$

Conversely, let $x \in (\limsup A_n) \setminus (\liminf A_n)$. Then (i) x is an element of infinitely many of the events $\{A_n : n = 1, 2, 3, \ldots\}$, and, (ii) x is not an element of all but finitely many of the events $\{A_n : n = 1, 2, 3, \ldots\}$. So (i) x is an element of infinitely many of the events $\{A_n : n = 1, 2, 3, \ldots\}$, and, (ii) x is an element of infinitely many of the events $\{A'_n : n = 1, 2, 3, \ldots\}$. Thus, there are two, infinite disjoint sets $\{j_1, j_2, j_3, \ldots\}$ and $\{k_1, k_2, k_3, \ldots\}$ such that

- x belongs to each of $\{A_{j_1}, A_{j_2}, A_{j_3}, \ldots\}$;
- x belongs to each of $\{A'_{j_1}, A'_{j_2}, A'_{j_3}, \ldots\}$;
- $\{j_1, j_2, \ldots\} \cup \{k_1, k_2, \ldots\} = \{1, 2, 3, \ldots\}$ (since, for each n, $x \in A_n$ or $x \in A'_n$).

It must happen infinitely often that $|j_n - k_m| = 1$. For suppose that $j_1 = 1$. Let

$$n_1 = \min\{p : j_{p+1} - j_p > 1\}.$$

Let $m_1 = 1$. Then $|j_{n_1} - k_{m_1}| = 1$. Let

$$m_2 = \min\{p : p > m_1; k_{p+1} - k_p > 1\}.$$

Let $n_2 = n_1 + 1$. Then $|j_{n_2} - k_{m_2}| = 1$. We keep this up *ad infinitum*. Hence, it must happen infinitely often that

$$x \in A_n \triangle A_{n+1};$$

that is, $x \in \limsup(A_n \triangle A_{n+1})$. Hence

$$(\limsup A_n) \setminus (\liminf A_n) \subset \limsup(A_n \triangle A_{n+1}).$$

vi. $\limsup A_n = \cap_{j=1}^{\infty} \cup_{i=j}^{\infty} A_i$. Let

$$B_j = \cup_{i=j}^{\infty} A_i \quad (j = 1, 2, 3, \ldots).$$

Then $B_1 \supset B_2 \supset B_3 \supset \ldots$ and

$$\limsup A_n = \cap_{j=1}^{\infty} B_j.$$

Now observe that

$$\begin{aligned} A_j &\subset B_j \\ P(A_j) &\leq P(B_j) \leq 1 \\ \limsup P(A_j) &\leq \limsup P(B_j) \\ &= \lim_{j\to\infty} P(B_j) \\ &= P(\lim B_j) \\ &= P(\cap_{j=1}^{\infty} B_j) \\ &= P(\limsup A_j). \end{aligned}$$

Similarly

$$\liminf P(A_j) \geq P(\liminf A_j).$$

(17) We will see that conditional probabilties are just ordinary probabilties with a particular measure function. Conditional probabilities

emphasise the importance of knowing the underlying measure in any probability problem.

(a) It suffices to prove that P_B is a probability measure. Observe the following.

- $P_B(\Omega) = P(\Omega \cap B)/P(B) = 1$.
- If $\{A_1, A_2, \ldots\}$ is a sequence of disjoint events, then

$$\begin{aligned} P_B(\cup_{i=1}^{\infty} A_i) &= P(B \cap \cup_{i=1}^{\infty} A_i)/P(B) \\ &= P(\cup_{i=1}^{\infty}(B \cap A_i))/P(B) \\ &= \sum_{i=1}^{\infty} P(B \cap A_i)/P(B) \\ &= \sum_{i=1}^{\infty} P_B(A_i). \end{aligned}$$

(b) Proof of Bayes' theorem.

$$\begin{aligned} P(X) &= P(X \cap \Omega) \\ &= P(X \cap \cup_{i=1}^{n} Y_i) \\ &= P(\cup_{i=1}^{n}(X \cap Y_i)) \\ &= \sum_{i=1}^{n} P(X \cap Y_i) \\ &= \sum_{i=1}^{n} P(Y_i) P_{Y_i}(X) \\ &= \sum_{i=1}^{n} P(Y_i) P(X|Y_i). \end{aligned}$$

So

$$\begin{aligned} P(Y_k|X) &= \frac{P(Y_k \cap X)}{P(X)} \\ &= \frac{P(Y_k \cap X)}{\sum_{i=1}^{n} P(Y_i)P(X|Y_i)} \\ &= \frac{P(Y_k)P(X|Y_k)}{\sum_{i=1}^{n} P(Y_i)P(X|Y_i)}. \end{aligned}$$

(18) Considering this sort of question can lead to reading about the connections between probability and psychology; more precisely

between the ways in which mathematicians measure risk and the general perceptions of risk.

(a) Without any calculation one may be tempted to argue as follows.

> In a court of law, the central issue is the reliability of the witness — not the distribution of black and white taxis in the city. As it has been proved that the witness is 80% reliable, we would tend to believe the witness. Hence, it is more likely that the taxi is white.

Below we see the intricate connections between the answer and the parameters of the question.

(b) We use Bayes' theorem. Define the following events.

Y_1 = taxi in accident is white.
Y_2 = taxi in accident is black.
X = witness says that taxi is white.

Thus, $P(Y_1) = 0.15$; $P(Y_2) = 0.85$; $P(X|Y_1) = 0.8$; $P(X|Y_2) = 0.2$. By Bayes' theorem

$$\begin{aligned} P(Y_1|X) &= \frac{P(Y_k)P(X|Y_k)}{\sum_{i=1}^{n} P(Y_i)P(X|Y_i)} \\ &= \frac{(0.15)(0.8)}{(0.15)(0.8) + (0.85)(0.2)} \\ &= 0.41; \end{aligned}$$

and similarly

$$P(Y_2|X) = 0.59.$$

(c) Hence it is more likely that the cab is black. This illustrates the risks associated with "wooly" thinking about probability.

(d) Here,

$$P(Y_1) = p;\ P(Y_2) = 1 - p;\ P(X|Y_1) = 0.8;\ P(X|Y_2) = 0.2.$$

By Bayes' theorem

$$P(Y_1|X) = \frac{P(Y_1)P(X|Y_1)}{\sum_{i=1}^{n} P(Y_i)P(X|Y_i)}$$

$$
\begin{aligned}
&= \frac{(p)(0.8)}{(p)(0.8)+(1-p)(0.2)} \\
&= \frac{0.8p}{0.2+0.6p}.
\end{aligned}
$$

So $P(Y_1|X) > 0.5$ if and only if $p > 0.2$.

(e) Here,

$$P(Y_1) = p;\ P(Y_2) = 1-p;\ P(X|Y_1) = q;\ P(X|Y_2) = 1-q.$$

By Bayes' theorem

$$
\begin{aligned}
P(Y_1|X) &= \frac{P(Y_1)P(X|Y_1)}{\sum_{i=1}^{n} P(Y_i)P(X|Y_i)} \\
&= \frac{pq}{pq+(1-p)(1-q)} \\
&= \frac{pq}{1-p-q+2pq}.
\end{aligned}
$$

Thus $P(Y_1|X) > 0.5$ if and only if

$$
\begin{aligned}
\frac{pq}{1-p-q+2pq} &> \frac{1}{2} \\
2pq &> 1-p-q+2pq \\
p+q &> 1.
\end{aligned}
$$

Chapter 2

Elementary probability

(1) A table of this type is often called a "contingency table" as it depicts all possible contingencies of the survey outcomes.

(a) $P(\text{intends or undecided}) = (29 + 31)/100 = 0.6$.
(b) $P(\text{age} \leq 40) = (17 + 37)/100 = 0.54$.
(c) $P(\text{age} \leq 25 \text{ and intends}) = 2/100 = 0.02$.
(d) $P(\text{age} \leq 25 \text{ or intends}) = (17 + 29 - 2)/100 = 0.4400$.
(e) $P(\text{intends} \mid \text{age } \leq 25) = 2/17 = 0.1176$.
(f) $P(\text{age} \leq 25 | \text{ intends}) = 2/29 = 0.0690$.
(g) Here $\wedge$ means "and", $\vee$ means "or", and $\neg$ means "not".

$$\begin{aligned}
& P(\text{age} \not> 40 \wedge \text{not undecided }) \\
= \; & P((\neg(\text{age} > 40)) \wedge (\neg(\text{undecided}))) \\
= \; & P(\neg((\text{age} > 40) \vee (\text{undecided}))) \\
= \; & 1 - P((\text{age} > 40) \vee (\text{undecided})) \\
= \; & 1 - (P(\text{age} > 40) + P(\text{undecided}) \\
& \quad -P((\text{age} > 40) \wedge (\text{undecided}))) \\
= \; & 1 - (46 + 31 - 16)/100 \\
= \; & 0.39.
\end{aligned}$$

(h) More use of logic notation.

$$\begin{aligned}
& P(\text{age} \not> 40 \vee \text{not undecided}) \\
= \; & P(\neg(\text{age} > 40) \vee \neg(\text{undecided}))
\end{aligned}$$

$$\begin{aligned} &= P(\neg(\text{age} > 40 \wedge \text{undecided})) \\ &= 1 - P(\text{age} > 40 \wedge \text{undecided}) \\ &= 1 - (16/100) \\ &= 0.84. \end{aligned}$$

(2) The study of the distribution of last digits in tables is an interesting source of problems.

(a) P(The missing numbers are 7, 4, 0 (in this order)) $= (0.1)^3 = 0.001$.
(b) P(The set of missing numbers is $\{0, 4, 7\}$) $= 3!(0.1)^3 = 0.006$.
(c) P(The missing digits are all equal to each other) $= \binom{10}{1}(0.1)^3 = 0.01$.
(d) First we determine the number of ways in which two of the missing digits are equal to each other, but the third is different from the other two. We

- choose the missing digit of which there will be two
- choose the other missing digit
- decide the order of the three digits.

Thus the number of ways $= (10)(9)(3) = 270$ and the required probability is 0.270.
(e) First we determine the number of ways in which the three missing digits are different from each other. We

- choose three distinct digits,
- decide the order of the three digits.

Thus the number of ways $= \binom{10}{3}6 = 720$ and the required probability is 0.720.
(f) The probabilities in the last three questions sum to 1 because they represent the probabilities of three exclusive and exhaustive events.

(3) There is a technical problem associated with how you choose a random sample of telephone numbers from a telephone directory. You must devise a means by which every number in the book has the same chance of being selected. The problem is important in conducting sample surveys by telephone. For our purposes, it suffices to note this problem rather than solve it.

We illustrate the analysis of such data by considering data from the Bendigo telephone directory. A description of the χ^2 goodness-of-fit test can be found in many standard books on statistical analysis. Let X denote the last digit of a telephone number chosen at random from the Bendigo telephone directory. We are testing the null hypothesis that X is uniformly distributed over the set

$$\{0, 1, 2, 3, 4, 5, 6, 7, 8, 9\}$$

against the alternative that it has some other distribution. Our data collected can be summarised as follows.

X	0	1	2	3	4	5	6	7	8	9
Frequency	24	23	27	20	26	14	20	14	15	17

Thus, $\chi^2 = 11.15$, d.f. $= 9$, and

$$P(\chi^2(9) > 11.15) = 0.2656.$$

(This can be calculated from spreadsheet packages.) So the sample data is consistent with the null hypothesis.

(4) A χ^2 goodness-of-fit test could be applied to test the null hypothesis that the numbers chosen are uniformly distributed over the set

$$\{0, 1, 2, \ldots, 45\}$$

against the alternative that it has some other distribution. This analysis leads to the statistic $\chi^2 = 31.77$, d.f. $= 44$, and

$$P(\chi^2(44) > 31.77) = 0.9156.$$

(This can be calculated from spreadsheet packages.) So the sample data is consistent with the null hypothesis.

(5) In each case, the probability is

$$\frac{1}{\binom{45}{6}} = 1.2277 \times 10^{-7}$$

because all selections are equally likely. I suspect that many people may choose "unusual" combinations such as

$$\{1, 2, 3, 4, 5, 6\}$$

believing that this combination is so unlikely that no one would choose it and hence they could win lots of money. However, if many people think this way, then when that combination is the winning set, there will be lots of winners and the prize for each winner may be quite small.

(6) This example suggests that there are interesting connections between probability theory and number theory. In fact, there is a branch of mathematics known as "probabilistic number theory". To answer this specific question, note that

$$\begin{aligned}
& X^2-1 \text{ is divisible by } 10 \\
\Leftrightarrow \quad & \text{The decimal expression for } X^2-1 \text{ ends in zero} \\
\Leftrightarrow \quad & \text{The decimal expression for } X^2 \text{ ends in } 1 \\
\Leftrightarrow \quad & \text{The decimal expression for } X \text{ ends in 1 or 9.}
\end{aligned}$$

So $p(n) =$ (the number of elements of $\{1, 2, \ldots, n\}$ with decimal expressions which end in 1 or 9)$/n$. Thus $p(10) = 2/10 = 0.2$; $p(25) = 5/25 = 0.2$. A little reflection leads to

$$p(n) = \frac{2\left[\frac{n}{10}\right] + \epsilon_n}{n}$$

where $\epsilon_n = 1$ or 2. Thus

$$\lim_{n\to\infty} p(n) = 0.2.$$

(7) In a game of $6n$ throws, let

$$X_n = \#\text{ times a score is 1.}$$

$$\begin{aligned}
& P(\text{Win Game 1}) \\
= \quad & P(X_1 \geq 1) \\
= \quad & 1 - P(X_1 = 0) \\
= \quad & 1 - \left(\frac{5}{6}\right)^6 \\
= \quad & 0.6651.
\end{aligned}$$

$$\begin{aligned}
& P(\text{Win Game 2}) \\
= \; & P(X_2 \geq 2) \\
= \; & 1 - P(X_2 = 0) - P(X_2 = 1) \\
= \; & 1 - \left(\frac{5}{6}\right)^6 - \binom{12}{1}\left(\frac{5}{6}\right)^{11}\left(\frac{1}{6}\right) \\
= \; & 0.6187.
\end{aligned}$$

$$\begin{aligned}
& P(\text{Win Game 3}) \\
= \; & P(X_3 \geq 3) \\
= \; & 1 - P(X_3 = 0) - P(X_3 = 1) - P(X_3 = 2) \\
= \; & 1 - \left(\frac{5}{6}\right)^{18} - \binom{18}{1}\left(\frac{5}{6}\right)^{17}\left(\frac{1}{6}\right) \\
& \quad - \binom{18}{2}\left(\frac{5}{6}\right)^{16}\left(\frac{1}{6}\right)^2 \\
= \; & 0.5973.
\end{aligned}$$

Thus, you are most likely to win Game 1.

(8) The aim of this exercise is to demonstrate the use of tree diagrams.

(a) The tree diagram is shown below.

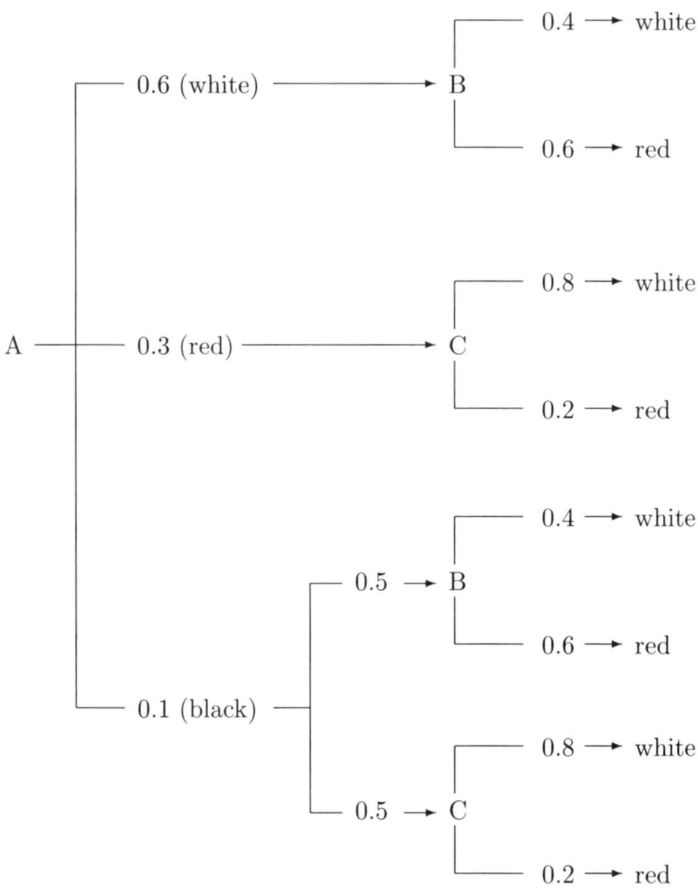
A
0.6 (white)
0.3 (red)
0.1 (black)
B
C
0.5
0.5
0.4
white
0.6
red
0.8
white
0.2
red

(b) P(First ball chosen from A was white) $= 0.6$.

(c) P(Ball placed in D is red)
$= (0.6)(0.6) + (0.3)(0.2) + (0.1)(0.5)(0.6) + (0.1)(0.5)(0.2)$
$= 0.46$.

(d) This is a problem in conditional probability. Let Aw denote the event that the first ball chosen from A was white. Let Dr denote the event that the ball placed in D was red. Then

$$\begin{aligned} P(Aw|Dr) &= \frac{P(Aw \textbf{ and } Dr)}{P(Dr)} \\ &= \frac{0.36}{0.46} \\ &= 0.7826. \end{aligned}$$

(e) Knowing that the ball placed in D is red has altered the likelihood that the first ball chosen from A was white. This is in fact an application of Bayes' theorem. However, using the tree diagram makes the argument more transparent than trying to use the formula in Bayes' theorem.

(9) This is the classical birthday problem.

(a) The probability that all n birthdays are different is given by:

$$1 - p(n) = \frac{365}{365} \cdot \frac{364}{365} \cdot \frac{363}{365} \cdot \ldots \cdot \frac{365 - n + 1}{365}.$$

Thus

$$p(n) = 1 - \frac{365}{365} \cdot \frac{364}{365} \cdot \frac{363}{365} \cdot \ldots \cdot \frac{365 - n + 1}{365}.$$

(b) Calculations of $p(n)$ can be easily made on a spreadsheet.

n	10	20	30	40
$p(n)$	0.1169	0.4114	0.7063	0.8912

(c) From the above expression for $p(n)$, it follows that if n increases, then $p(n)$ increases. Also $p(23) = 0.5073$. Thus, if $n \geq 23$ then $p(n) > 0.5$.

(d) Thus, if we have at least 23 persons in a room, it is more likely than not that at least 2 persons have their birthdays on the same day.

(e) $p(41) = 0.9032$. Thus, if $n \geq 41$ then $p(n) > 0.9$.

(10) This is another birthday problem.

(a) P(All n birthdays are not 29 May) $= (364/365)^n$. Hence

$$q(n) = 1 - (364/365)^n.$$

(b) Calculations of $q(n)$ can be easily made on a spreadsheet.

n	10	20	30
$q(n)$	0.0271	0.0534	0.0790

(c) Solve the equation $q(n) = p$ for n in terms of p as follows.

$$\begin{aligned} q(n) &= p \\ 1 - (364/365)^n &= p \\ (364/365)^n &= 1 - p \\ n \ln(364/365) &= \ln(1-p) \\ n &= -364.5 \ln(1-p). \end{aligned}$$

(d) From the above expression for $q(n)$, it follows that if n increases, then $q(n)$ increases. Also $q(253) = 0.5005$. Thus, if $n \geq 253$ then $q(n) > 0.5$.

(e) If there are at least 253 people in the room, then it is more likely than not that someone will have their birthday on 29 May.

(f) By above, $q(n) > 0.9$ if $n > -364.5 \ln(1 - 0.9) = 839.29$. So, if $n > 840$ then $q(n) > 0.9$.

(11) Let A_k denote A's score on the k-th turn; B_k denote B's score on the k-th turn. Then $P(A_k = 6) = 5/36$ and $P(B_k = 7) = 6/36$.

(a) Then

$$\begin{aligned} & p_A(n) \\ = \; & P(A_1 = 6) + P(A_1 \neq 6, B_1 \neq 7, A_2 = 6) + \ldots \\ & + P(A_1 \neq 6, B_1 \neq 7, \ldots, A_{n-1} \neq 6, B_{n-1} \neq 7, A_n = 6) \\ = \; & \sum_{k=0}^{n-1} \left(\frac{31}{36}\right)^k \left(\frac{30}{36}\right)^k \left(\frac{5}{36}\right) \end{aligned}$$

$$= \frac{5}{36}\left(\frac{1-\left(\frac{31}{36}\right)^n\left(\frac{30}{36}\right)^n}{1-\left(\frac{31}{36}\right)\left(\frac{30}{36}\right)}\right)$$
$$= (0.4918)(1-(0.7176)^n).$$

(b) Also,

$$\begin{aligned} & p_B(n) \\ = \; & P(A_1 \neq 6, B_1 = 7) \\ & +P(A_1 \neq 6, B_1 \neq 7, A_2 \neq 6, B_2 = 7) \\ & +\dots \\ & +P(A_1 \neq 6, B_1 \neq 7, \\ & \quad A_2 \neq 6, B_2 \neq 7, \\ & \quad \dots \\ & \quad A_n \neq 6, B_n = 7) \\ = \; & \sum_{k=0}^{n}\left(\frac{31}{36}\right)^{k+1}\left(\frac{30}{36}\right)^k\left(\frac{6}{36}\right) \\ = \; & \left(\frac{31}{36}\right)\left(\frac{6}{36}\right)\left(\frac{1-\left(\frac{31}{36}\right)^{n+1}\left(\frac{30}{36}\right)^{n+1}}{1-\left(\frac{31}{36}\right)\left(\frac{30}{36}\right)}\right) \\ = \; & (0.5082)(1-(0.7176)^{n+1}). \end{aligned}$$

(c) If, in n turns, A does not win and B does not win then the game is drawn. Hence

$$d(n) = 1 - p_A(n) - p_B(n) = (0.8565)(0.7176)^n.$$

(d) Thus,

$$\begin{aligned} \lim_{n\to\infty} p_A(n) &= 0.4918, \\ \lim_{n\to\infty} p_B(n) &= 0.5082, \\ \lim_{n\to\infty} d(n) &= 0. \end{aligned}$$

(12) Life tables are useful for demonstrating probability concepts. For further reading see Namboodiri and Suchindran [49].

(a) Expressions required are as follows.

- $d(x) = \ell(x) - \ell(x+1)$.
- $p(x) = \ell(x+1)/\ell(x)$
- $q(x) = (\ell(x) - \ell(x+1))/\ell(x)$

(b) Required probabilities are as follows.

i. P(Male survives to age 30)
= 96,559/100,000 = 0.96559.

ii. P(Male survives to age 20 but does not reach age 60)
= (97,956-84,782)/100,000 = 0.13174.

iii. P(Male does not survive to age 60 | He has survived to age 20)
= (97,956-84,782)/97,956 = 0.13449.

iv. P(Male does not survive to age 55)
= 1 - 0.89566 = 0.10434.

(c) Use official statistics to estimate the age-specific death rate ($q(x)$) for each age (x). This is possible if we know the number of males in the population aged x and the number of males who died aged x. Then, starting with an arbitrary $\ell(0)$ (say 100,000) the rest of the table unfolds from

$$\ell(x+1) = \ell(x)(1 - q(x)).$$

(13) Points to note on the graphs are the following.

- The mortality rates in the first few years are relatively high. The level of infant mortality is an important national indicator of health care.
- Mortality rates for males are consistently higher than those for females. This makes sense for many ages (e.g. 15–25); but why is this so for ages 0–2?
- Mortality rates rise steeply for ages 15–20.

If you like problem solving, probability and demography look at the book of problems by N. Keyfitz and J. Beckman [39].

(14) It is interesting the note that the probability of getting none of the regular patterns (straight flush, ..., 1 pair) just exceeds 0.5. Whoever designed this game was pretty clever.

(a) Number of possible Poker hands $= \binom{52}{5} = 2,598,960$.

(b) There is an art to setting out answers to questions such as this because the answers need a greater emphasis on words than on symbols. Below we set out the arguments for calculating the number of hands; the probability of the hand is found by dividing the number of hands by 2,598,960. Sometimes one can choose from a variety of arguments and we have illustrated this in the different parts.

Straight flush. To construct a flush proceed as follows. Choose one of the 4 suits and then choose one of the 10 possible sequences. Thus the number of flushes is $(4)(10) = 40$.

Four of a kind. To construct such a hand, proceed as follows. First choose one of the 13 kinds; we will use all of this kind in our hand; then choose the odd card out of the remaining 48. Thus, the number of hands is $(13)(48) = 624$.

Full house. This looks like $\{X, X, X, Y, Y\}$ where X and Y are different face values. To construct a full house, proceed as follows. Choose 2 out of 13 face values; this can be done in $\binom{13}{2}$ ways. We decide which one will be X; this can be done in $\binom{2}{1}$ ways. We choose 3 of the possible 4 X cards; this can be done in $\binom{4}{3}$ ways. Finally we choose 2 of the possible 4 Y cards; this can be done in $\binom{4}{2}$ ways. Thus the number of full house hands is

$$\binom{13}{2} \cdot \binom{2}{1} \cdot \binom{4}{3} \cdot \binom{4}{2} = 3744.$$

Flush. To construct such a hand, proceed as follows. Choose the suit and then choose 5 from that suit not in sequence. Thus the total number of hands is

$$\binom{4}{1}\left(\binom{13}{5} - 10\right) = 5108.$$

Straight. There are 10 possible sequences; choose one. Then for each element of the sequence we have a choice of 4 suits. However, we do not want to include any of the 40 straight flushes. Thus, the number of hands is

$$(10.4.4.4.4.4) - 40 = 10,200.$$

Three of a kind. Choose the face value of which there will be three; choose 3 of the 4 cards with this face value; finally choose 2 cards from the 48 with different face values. Thus, the total number of hands is

$$\binom{13}{1}\binom{4}{3}\left(\binom{48}{2} - \binom{12}{1}\binom{4}{2}\right) = 54,912.$$

Two pair. This looks like $\{X, X, Y, Y, Z\}$ where X, Y and Z are different face values. To construct such a hand, proceed as follows.

- Choose 2 out of 13 face values to be X and Y; this can be done in $\binom{13}{2}$ ways.
- Choose 2 of the X cards; this can be done in $\binom{4}{2}$ ways.
- Choose 2 of the Y cards; this can be done in $\binom{4}{2}$ ways.
- Choose 1 out of 44 cards which are neither X nor Y; this can be done in $\binom{44}{1}$ ways.

Thus, the total number of hands is

$$\binom{13}{2}\binom{4}{2}\binom{4}{2}\binom{44}{1}1 = 123,552.$$

One pair. This looks like $\{X, X, U, V, W\}$. To construct such a hand, proceed as follows.

- Choose 1 out of 13 face values to be X; this can be done in $\binom{13}{1}$ ways.

- Choose 2 of the X cards; this can be done in $\binom{4}{2}$ ways.
- Choose the 3 remaining face values to be U, V, W; this can be done in $\binom{12}{3}$ ways.
- For each of the 3 face values U, V, W choose one of the 4 possible suits; this can be done in 4^3 ways.

Thus, the total number of hands is

$$\binom{13}{1}\binom{4}{2}\binom{12}{3}4^3 = 1,098,240.$$

None of the above. The balance of hands are in this category.

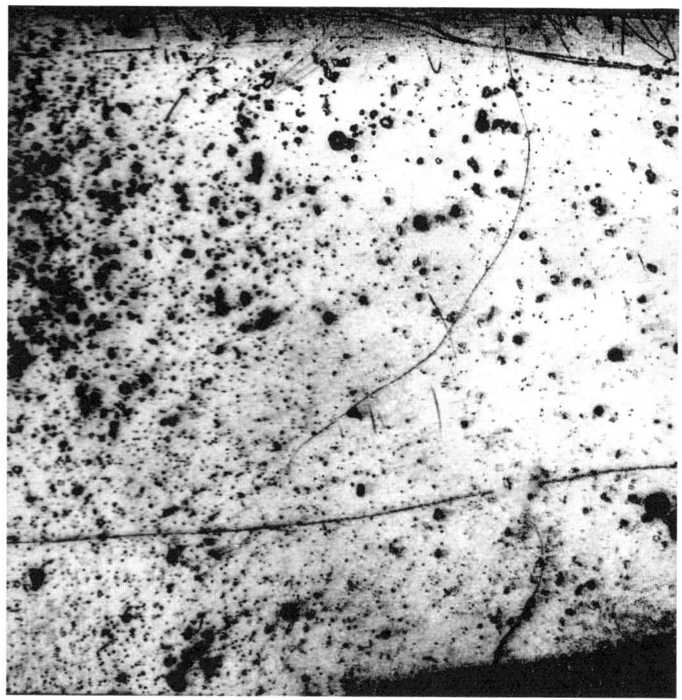

Chapter 3

Discrete random variables

(1) Let $\zeta(s) = \sum_{t=1}^{\infty} t^{-s}$ for $s > 1$. Note that for $s \leq 1$, this series is divergent.

(a) Suppose that the distribution of X is given by

$$P(X = t) = (\zeta(2)t^2)^{-1} \quad (t = 1, 2, 3, \ldots).$$

This is a *bona fide* distribution because $P(X = t) > 0$ for $t = 1, 2, 3, \ldots$ and $\sum_{t=1}^{\infty} P(X = t) = 1$. However

$$\sum_{t=1}^{\infty} tP(X = t) = \zeta(2)^{-1} \sum_{t=1}^{\infty} t^{-1}$$

and hence μ_X are not defined.

(b) Suppose that the distribution of X is given by

$$P(X = t) = (C(3)t^3)^{-1} \quad (t = 1, 2, 3, \ldots).$$

Then

$$\mu_X = \sum_{t=1}^{\infty} tP(X = t) = C(2)/C(3);$$

however

$$\sum_{t=1}^{\infty} t^2 P(X = t) = \zeta(3)^{-1} \sum_{t=1}^{\infty} t^{-1}$$

and hence $E(X^2)$, and therefore σ_X, are not defined.

(c) Suppose that the distribution of X is given by

$$P(X = t) = (\zeta(p+2)t^3)^{-1} \quad (t = 1, 2, 3, \ldots).$$

Then, for $1 \le k \le p$,

$$E(X^k) = \sum_{t=1}^{\infty} tP(X = t) = \zeta(p + 2 - k)/\zeta(p + 2);$$

however, for $k \ge p + 1$, $p + 2 - k \le 1$, it follows that

$$\sum_{t=1}^{\infty} t^k P(X = t) = \zeta(p + 2)^{-1} \sum_{t=1}^{\infty} t^{p+2-k}$$

and hence $E(X^k)$ are not defined.

(2) This example concerns a portfolio with only two investments. Imagine how complex the matter becomes for a portfolio with many investments.

(a) A = 0.09; B = 0.11; C = 0.35; D = 0.12

(b) The required probabilities are:

i. $P(X = 0 \text{ and } Y = 2) = 0.11$
ii. $P(X = 0 \text{ and } Y \le 1.5) = 0.19$
iii. $P(X = 0 \text{ or } Y = 3) = 0.44$
iv. $P(X = 2) = 0.29$
v. $P(X \ne 2 \text{ and } Y = -1) = 0.08$
vi. $P(X < 0 \text{ or } Y \ne 1) = 0.78$

(c) We have $E(XY) = \sum stP(X = s, Y = t) = -0.06$.

(d) The distribution of X is given by the following table.

s	-2	0	2	10
$P(X = s)$	0.24	0.35	0.29	0.12

(e) $E(X) = \mu_X = \sum sP(X = s) = 1.3$.
$E(X^2) = \sum s^2 P(X = s) = 14.12$.
$\sigma_X^2 = E(X^2) - E(X)^2 = 12.43 = (3.526)^2$.

(f) The distribution of Y is given by the following table.

t	-1	0	1	2	3
$P(Y = t)$	0.13	0.23	0.30	0.20	0.14

(g) $E(Y) = \mu_Y = \sum tP(Y = t) = 0.99$.
$E(Y^2) = \sum t^2 P(Y = t) = 2.49$.
$\sigma_Y^2 = E(Y^2) - E(Y)^2 = 1.5099 = (1.229)^2$.

(h) The coefficient of correlation between X and Y is

$$\begin{aligned} \rho_{X,Y} &= \frac{E(XY) - E(X)E(Y)}{\sigma_X \sigma_Y} \\ &= \frac{-0.06 - (1.3)(0.99}{(3.526)(1.229)} \\ &= -0.31. \end{aligned}$$

(i) The entries in Table 3.1 suggest that low values of X (e.g. $s = -2$) tend to be clustered around higher values of Y (e.g. around $t = 1, 2, 3$). Thus the coefficient of correlation between X and Y ought to be negative.

(3) We have defined $XY = 10\xi_1 + \xi_2$.

(a) Values of (ξ_1, ξ_2) for all possible values (X, Y) are shown in the two tables below.

Table 3.1 Table of values of (ξ_1, ξ_2)

		X				
		0	1	2	3	4
	0	(0,0)	(0,0)	(0,0)	(0,0)	(0,0)
	1	(0,0)	(0,1)	(0,2)	(0,3)	(0,4)
	2	(0,0)	(0,2)	(0,4)	(0,6)	(0,8)
	3	(0,0)	(0,3)	(0,6)	(0,9)	(1,0)
Y	4	(0,0)	(0,4)	(0,8)	(1,2)	(1,6)
	5	(0,0)	(0,5)	(1,0)	(1,5)	(2,0)
	6	(0,0)	(0,6)	(1,2)	(1,8)	(2,4)
	7	(0,0)	(0,7)	(1,4)	(2,1)	(2,8)
	8	(0,0)	(0,8)	(1,6)	(2,4)	(3,2)
	9	(0,0)	(0,9)	(1,8)	(2,7)	(3,6)

Table 3.2 Table of values of (ξ_1, ξ_2)

		X				
		5	6	7	8	9
	0	(0,0)	(0,0)	(0,0)	(0,0)	(0,0)
	1	(0,5)	(0,6)	(0,7)	(0,8)	(0,9)
	2	(1,0)	(1,2)	(1,4)	(1,6)	(1,8)
	3	(1,2)	(1,5)	(1,8)	(2,1)	(2,4)
Y	4	(2,0)	(2,4)	(2,8)	(3,2)	(3,6)
	5	(2,5)	(3,0)	(3,5)	(4,0)	(4,5)
	6	(3,0)	(3,6)	(4,2)	(4,8)	(5,4)
	7	(3,5)	(4,2)	(4,9)	(5,6)	(6,3)
	8	(4,0)	(4,8)	(5,6)	(6,4)	(7,2)
	9	(4,5)	(5,4)	(6,3)	(7,2)	(8,1)

(b) The two tables below present the distribution of the vector (ξ_1, ξ_2).

Table 3.3 Distribution of (ξ_1, ξ_2); missing values are zero.

		ξ_2				
		0	1	2	3	4
	0	0.19	0.01	0.02	0.02	0.03
	1	0.02		0.04		0.02
	2	0.02	0.02			0.04
	3	0.02		0.02		
ξ_1	4	0.02		0.02		
	5					0.02
	6				0.02	0.01
	7			0.02		
	8		0.01	0.02		

Table 3.4 Distribution of (ξ_1, ξ_2); missing values are zero ctd.

		ξ_2				
		5	6	7	8	9
	0	0.02	0.04	0.02	0.04	0.03
	1	0.02	0.03		0.04	
	2	0.01		0.02	0.02	
	3	0.02	0.03			
ξ_1	4	0.02			0.02	0.01
	5		0.02			
	6					
	7					
	8					

(c) The most likely value of the vector (ξ_1, ξ_2) is $(0, 0)$; $P((\xi_1, \xi_2) = (0, 0)) = 0.19$.

(d) Here we use the distribution of (ξ_1, ξ_2) as presented in the two tables above. Let $S = \{(0,0), (0,1), (0,4), (0,9), (1,6), (2,5), (3,6), (4,9), (6,4), (8,1)\}$.
Then

$$P(XY \text{ is a perfect square}) = P((\xi_1, \xi_2) \in S) = 0.36.$$

(e) Observe that

$$\begin{aligned} P((\xi_1, \xi_2) = (0,0)) &= 0.19; \\ P(\xi_1 = 0) &= 0.42; \\ P(\xi_2 = 0) &= 0.27. \end{aligned}$$

So,

$$P((\xi_1, \xi_2) = (0,0)) \neq P(\xi_1 = 0) P(\xi_2 = 0).$$

Thus the random variables ξ_1 and ξ_2 are not independent.

(f) The distribution of ξ_1, is given by the following table.

t	0	1	2	3	4
$P(\xi_1 = t)$	0.42	0.17	0.13	0.09	0.09
t	5	6	7	8	
$P(\xi_1 = t)$	0.04	0.03	0.02	0.01	

Thus

$$\begin{aligned} E(\xi_1) &= \sum tP(\xi_1 = t) = 1.66 \\ E(\xi_1^2) &= \sum t^2 P(\xi_1 = t) = 6.64 \\ \text{Var}\ (\xi_1) &= E(\xi_1^2) - E(\xi_1)^2 = 1.66. \end{aligned}$$

(g) The distribution of ξ_2, is given by the following table.

t	0	1	2	3	4
$P(\xi_2 = t)$	0.27	0.04	0.12	0.04	0.12
t	5	6	7	8	9
$P(\xi_2 = t)$	0.09	0.12	0.04	0.12	0.04

Thus

$$\begin{aligned} E(\xi_2) &= \sum tP(\xi_2 = t) = 3.65 \\ E(\xi_2^2) &= \sum t^2 P(\xi_2 = t) = 22.25 \\ \text{Var}\ (\xi_2) &= E(\xi_2^2) - E(\xi_2)^2 = 8.93. \end{aligned}$$

(h) Finally, we obtain

$$\begin{aligned} E(\xi_1\xi_2) &= \sum_{t_1=0}^{8}\sum_{t_2=0}^{9} t_1 t_2 P(\xi_1 = t_1)P(\xi_2 = t_2) \\ &= 6.3; \\ \text{Cov}\ (\xi_1, \xi_2) &= E(\xi_1\xi_2) - E(\xi_1)E(\xi_2) \\ &= 6.3 - (1.66)(3.65) \\ &= 0.241; \\ \text{Corr}\ (\xi_1, \xi_2) &= \frac{\text{Cov}\ (\xi_1, \xi_2)}{\sigma_{\xi_1}\sigma_{\xi_2}} \\ &= \frac{0.241}{\sqrt{3.88}\sqrt{8.93}} \\ &= 0.0409. \end{aligned}$$

(4) In each case, we describe features of an abstract model and a practical situation which fits the model. This questions illustrates the power of abstract models to describe many different situations.

(a) The abstract binomial model can be described as follows. We conduct a trial which can result in one of two outcomes which we will label as "success" or "failure". Suppose that we conduct a series of n trials which are independent of each other; that is, the result of one trial is not effected by the result of any other trial. Suppose further that the probability that a trial results in "success" is constant from trial to trial; we denote this probability by p. Finally we let X denote the number of successes which occurred in the n trials. Then X has a binomial distribution with parameters n and p.

Here is a practical situation which fits this model.

We conduct a survey of people in Bendigo and ask them if they believe that Australia should become a republic. (This has been a hot issue in recent times in Australia.) We classify their responses as either "Yes" or "something other than Yes" (such as "No", "I don't have any opinion"). In this situation, the trial consists of choosing a person at random and asking them the question "Do you believe that Australia should become a republic?". The outcome is "Yes" or something else. We conduct the survey ensuring that responses from individuals are independent of each other; so, for example, we do not ask people in groups. We assume that p, the probability of "Yes" is an unknown but fixed constant by conducting the survey over a relatively brief period of time. We would not conduct the survey over a year because p may be changing in our population over this time. Then we let X denote the number of respondents who said "Yes"; X will have a binomial distribution with parameters n and p. We can control n by choosing an appropriate sample size (say $n = 200$) and this author imagines that p may be about 0.5 in Bendigo at the time of writing.

If we allowed more that two reponses to our question (such as "Yes", "No", "I have not decided", "Other") then the situation would lead to a multinomial distribution.

(b) General features of the Poisson model can be described as follows. First, X can assume only the values in the set

$$\{0, 1, 2, 3, \ldots\}.$$

Second, the Poisson distribution is often described as the distribution of "rare events": this means that the probability that X is large tends to be very small. Finally, the parameter λ is the value of both μ_X and σ_X^2.

Here is a practical situation which may fit this model. The number of suicides (S) recorded in a given town each year may follow a Poisson distribution. This variable will assume only values in $\{0, 1, 2, 3, \ldots\}$ and, fortunately, such tragic events are rare. Perhaps, in the particular town, $\lambda = \mu_S$ may be 9.

You may object along the lines that S could not exceed the total size of the population and hence there are some (infinitely many) values of

$$\{0, 1, 2, 3, \ldots\}$$

which S could not assume. This is true; this is just part of the modelling process in which we make various assumptions to make the mathematics more feasible. Other investigators can come along, remove the assumptions and try to develop more realistic models.

Perhaps we could regard S as a binomial r.v. as follows. Suppose that n represents the population of the town and that p represents the probability that a person in the town will commit suicide that year. Then S fits the binomial model described above. Again, you may be argued that this model is unsatisfactory because "all members of the population are not necessarily exposed to the same risk" [66, p. 182].

Is the distribution of S binomial or Poisson? This question is provided by a limit theorem which we discuss in Problem 2 Chapter 4. If X has a binomial distribution, n is large, and p is small and np is moderate, then the distribution of X may be approximated by a Poisson distribution. Thus, it may well be that the distribution of S is a binomial distribution which is also approximately equal to a Poisson distribution.

(c) Many games of chance lead to examples of r.vs with the uniform distribution. For example, if we throw a fair die and we denote the score which results from the throw by X then X is uniformly distributed over the set $\{1, 2, 3, 4, 5, 6\}$.

(Note that "die" is singular and "dice" is plural. When playing a game, we often say "throw the dice" when, in fact, we have only one die to throw. On the other hand, one often hears the phrase that "the die is cast". It is interesting that the words "die" and "data" which are both associated in different ways with probability and statistics, come from the same Latin root "datum".)

(d) The hypergeometric distribution arises naturally in quality control. Suppose that we have a batch of $N = 5000$ manufactured items and M of these items are defective. We choose a sample of $n = 100$ items from the batch. On inspection we find that X of the sampled items are defective. Then X has a hypergeometric distribution. State some sensible values of the a parameters N and n involved.

(5) A classic reference for the study of Monte Carlo methods is Hammersley and Handscomb [30].

(a) The square $ABCD$ where $A = (0,0)$, $B = (0,1)$, $C = (1,1)$ and $D = (1,0)$ will act like a dart board at which we throw random points.

(b) You can easily do this on a spreadsheet.

(c) The results of my experiment looked like this.

x	x^2	y	Is $y < x^2$? (Yes = 1, No = 0)
0.7948	0.6317	0.1061	1
0.6688	0.4473	0.1575	1
0.6270	0.3931	0.5704	0
0.0679	0.0046	0.7037	0
0.3315	0.1099	0.0422	1
0.4255	0.1810	0.6731	0
0.5085	0.2586	0.0158	1
0.2505	0.0628	0.0759	0
0.2266	0.0513	0.5364	0
$\vdots$	$\vdots$	$\vdots$	$\vdots$

(d) I obtained $U = 34$ points under the curve and hence $U/n = U/100 = 0.34$ is my estimate of I.

(e) Because the area of the enclosing rectangle $ABCD$ is 1, the value of I is the probability that the random point (x, y) falls under the curve. The classical binomial models applies as described below.

A trial consists of choosing a point at random in the rectangle; each trial results in success (point under curve) or failure (point not under curve); the n trials are independent of each other; the probability of success $p = I$ is constant. Thus, the number of successes (i.e. the number of points under the curve which we denote by U) has a binomial distribution with parameters $n = 100$ and $p = I$.

(f) The length of the arc of the graph $y = \sin x \quad (0 \leq x \leq \pi)$. is given by $I = \int_0^\pi \sqrt{1 + \cos^2 x}\, dx$. This integral cannot be evaluated in any closed form. Thus we resort to a numerical method such as the Monte Carlo method.

To evaluate this integral, draw a rectangle $ABCD$ where $A = (0, 0)$, $B = (0, \sqrt{2})$, $C = (\pi, \sqrt{2})$ and $D = (\pi, 0)$. This encloses the required area. Then we carry out the experiment with $n = 100$ as described above. Here is an extract from my output.

x	$y_1 = \sqrt{1 + \cos^2(x)}$	y	Is $y < y_1$? Yes = 1; No = 0
0.1253	1.4087	1.2748	1
0.6428	1.2809	0.0492	1
0.8339	1.2048	0.7580	1
1.5483	1.0003	0.3112	1
1.6217	1.0013	0.6215	1
1.1943	1.0655	1.3260	0
2.2312	1.1732	0.9171	1
1.7935	1.0241	0.2597	1
3.0838	1.4130	1.2826	1
⋮	⋮	⋮	⋮

This led to $n = 100$; number of points under curve was $U = 85$;

hence the estimated value of I is:

$$\hat{I} = \frac{85}{100} \cdot \pi\sqrt{2} = 3.7765.$$

This can be compared with 3.8202 which was found by using a numerical integration facility in `Maple`.

It is clear from this exercise that writing a program to conduct Monte Carlo calculations is much easier that writing a program to carry out more sophisticated numerical methods. Thus, it is not surprising that Monte Carlo methods really show their use in multivariate problems when the number of variables is huge.

There are many other ways in which Monte Carlo methods can be applied in computational mathematics. For a description of how to use probabilistic methods to find the inverse of a matrix, see Harward and Mills [31]

(6) Starting at the right place is the key to these exercises.

(a) We begin with an expression for $E(X^2)$.

$$\begin{aligned}
E(X^2) &= \sum t^2 P(X=t) \\
&= \sum_{|t|<c} t^2 P(X=t) + \sum_{|t|\geq c} t^2 P(X=t) \\
&\geq 0 + c^2 \sum_{|t|\geq c} P(X=t); \\
\sum_{|t|\geq c} P(X=t) &\leq c^{-2} E(X^2); \\
P(|X| \geq c) &\leq c^{-2} E(X^2).
\end{aligned}$$

(b) Apply the first part to the r.v. $Z = X - \mu_X$. So

$$\begin{aligned}
P(|Z| \geq c) &\leq c^{-2} E(Z^2); \\
P(|X - \mu_X| \geq c) &\leq c^{-2} \sigma_X^2.
\end{aligned}$$

(c) As suggested above, we start off with an expression for $E(\phi(|X|))$.

$$\begin{aligned}
M = E(\phi(|X|)) &= \sum \phi(|t|) P(X=t) \\
&= \sum_{|t|<c} \phi(|t|) P(X=t) + \sum_{|t|\geq c} \phi(|t|) P(X=t)
\end{aligned}$$

$$
\begin{aligned}
&\geq 0+\phi(c)\sum_{|t|\geq c} P(X=t); \\
\sum_{|t|\geq c} P(X=t) &\leq M/\phi(c); \\
P(|X|\geq c) &\leq M/\phi(c).
\end{aligned}
$$

(d) Put $\phi(t)=t^2$.

(e) The collected works of P.L. Chebyshev (1821–1894) can be found in [14]. Other interesting sources of information are [48] and the article [13] and the article by Butzer and Jongmans [11].

(7) These questions show us just how beautiful is the series expansion for the exponential function.

(a) We calculate the m.g.f. of X as follows.

$$
\begin{aligned}
M_X(t) &= E(\exp(Xt)) \\
&= \sum_{k=0}^{\infty} \exp(kt)P(X=k) \\
&= \sum_{k=0}^{\infty} \exp(kt)\lambda^k \exp(-\lambda)/k! \\
&= \exp(-\lambda)\sum_{k=0}^{\infty} \exp(kt)\lambda^k/k! \\
&= \exp(-\lambda)\sum_{k=0}^{\infty} (\exp(t)\lambda)^k/k! \\
&= \exp(-\lambda+\lambda\exp(t)) \\
&= \exp(\lambda(\exp(t)-1)).
\end{aligned}
$$

(b) Hence

$$
\begin{aligned}
E(X) &= M'_X(0) \\
&= \exp(\lambda(\exp(t)-1))\lambda\exp(t)|_{t=0} \\
&= \lambda; \\
E(X^2) &= M''_X(t)|_{t=0} \\
&= M'_X(t)\lambda\exp(t)+M_X(t)\lambda\exp(t)|_{t=0}
\end{aligned}
$$

$$
\begin{aligned}
&= \lambda + \lambda^2; \\
\mathrm{Var}(X) &= E(X^2) - E(X)^2 \\
&= \lambda + \lambda^2 - (\lambda)^2 \\
&= \lambda.
\end{aligned}
$$

(c) Note that

$$\frac{P(X = t+1)}{P(X = t)} = \frac{\exp(-\lambda)\lambda^{t+1}}{(t+1)!} \cdot \frac{t!}{\exp(-\lambda)\lambda^t} = \frac{\lambda}{t+1};$$

therefore, since X assumes only integer values,

$$P(X = t) \leq P(X = t+1) \Leftrightarrow t \leq \lambda - 1 \Leftrightarrow t \leq [\lambda] - 1.$$

(8) This is an interesting property of the Poisson distribution.

(a) To find the distribution of $X_1 + X_2$, observe that, for $t \in \{0, 1, 2, \ldots\}$,

$$
\begin{aligned}
& P(X_1 + X_2 = t) \\
&= \sum_{k=0}^{t} P(X_1 = k, X_2 = t-k) \\
&= \sum_{k=0}^{t} P(X_1 = k)P(X_2 = t-k) \\
& \quad \text{(by independence of } X_1 \text{ and } X_2\text{)} \\
&= \sum_{k=0}^{t} \exp(-\lambda_1)\frac{\lambda_1^k}{k!} \cdot \exp(-\lambda_2)\frac{\lambda_2^{t-k}}{(t-k)!} \\
&= \frac{\exp(-(\lambda_1 + \lambda_2))}{t!} \sum_{k=0}^{t} \binom{t}{k} \lambda_1^k \lambda_2^{t-k} \\
&= \frac{\exp(-(\lambda_1 + \lambda_2))}{t!} (\lambda_1 + \lambda_2)^t \\
&= \frac{(\lambda_1 + \lambda_2)^t}{t!} \exp(-(\lambda_1 + \lambda_2)).
\end{aligned}
$$

So, $X_1 + X_2 \sim$ Poisson $(\lambda_1 + \lambda_2)$.

(b) Use induction and the previous result to show that if

$$X_i \sim \mathrm{Poisson}(\lambda_i)(i = 1, 2, \ldots, n)$$

and these rvs are independent of each other then

$$\sum_{i=1}^{n} X_i \sim \text{Poisson}(\sum_{i=1}^{n} \lambda_i).$$

(9) Again we encounter the link between the binomial distribution and the Poisson distribution.

(a) Here we have a classical binomial model. A trial consists of examining a note and classifying it as counterfeit or not; we have conducted $n = 100$ trials which are independent of each other; the probability of a note being counterfeit is constantly equal to $p = 150/20,000 = 0.0075$; X is the number of counterfeit notes in the sample. Thus, $X \sim$ Binomial$(n = 100, p = 0.0075)$.

(b) Using the formula for the binomial distribution, we obtain

$$\begin{aligned} P(X \geq 2) &= 1 - P(X = 0) - P(X = 1) \\ &= 1 - (1 - p)^n - np(1 - p)^{n-1} \\ &= 0.1730. \end{aligned}$$

(c) Here we offer a heuristic form of reasoning: a rigorous argument follows from Problem 2 in Chapter 4.2. The values which X can assume are $\{0, 1, 2, \ldots, 100\}$ and the average value of X is $np = 0.75$. So it may be reasonable to model X by $X \sim$ Poisson$(\lambda = 0.75)$ (approximately).

(d) Using this Poisson approximation we obtain

$$\begin{aligned} P(X \geq 2) &= 1 - P(X = 0) - P(X = 1) \\ &= 1 - \exp(-\lambda) - \lambda \exp(-\lambda) \\ &= 0.1734. \end{aligned}$$

(e) The Poisson approximation is easier to use and it seems to be reasonably accurate.

(10) This problem deals with "acceptance sampling" in statistical quality control. A useful reference for mathematical aspects of the subject is the text by A. Hald [28]. Let X denote the number of non-conforming cans in the sample. Then X is a r.v. with a

binomial distribution with parameters $n = 15$ and p. We accept the batch if $X \leq 2$ and so, $Pa = P(X \leq 2)$.

(a) To find an expression for Pa in terms of p, we proceed as follows.

$$\begin{aligned} Pa &= P(X \leq 2) \\ &= P(X = 0) + P(X = 1) + P(X = 2) \\ &= \binom{15}{0} p^0 (1-p)^{15} + \binom{15}{1} p^1 (1-p)^{14} \\ &\quad + \binom{15}{2} p^2 (1-p)^{13} \\ &= (1-p)^{15} + 15p(1-p)^{14} + 105p^2(1-p)^{13} \\ &= (1-p)^{13}(1 + 13p + 91p^2). \end{aligned}$$

(b) Points for the O.C. can be found by calculation. The O.C. is a summary of the effectiveness of the sampling plan. For any set level of quality p, the O.C. tells us the probability of batches of this quality passing through the system.

p	0.00	0.05	0.06	0.10	0.15	0.20	0.25	0.30
Pa	1.00	0.96	0.94	0.81	0.60	0.40	0.24	0.13

(c) From the above table, we see that if $Pa = 0.95$ then $0.05 < p < 0.06$. More detailed calculation leads to $p = 0.0565$ approximately. The value of p for which $Pa = 0.95$ is often referred to as the acceptable quality level (AQL).

(11) We continue our study of acceptance sampling.

(a) By knowing that the batches were "very large" we can assume that we are sampling without replacement (at least approximately) and this allows us to use the binomial distribution.

(b) Denote the batch size by $N = 100$ and since $p = 0.05$ we denote the number of non-conforming elements in the batch by $M = Np = 5$; the sample size is $n = 15$; we denote the number of non-conforming elements in the sample by the r.v. X. Then X has a hypergeometric distribution.

$$P(X = t) = \frac{\binom{M}{t}\binom{N-M}{n-t}}{\binom{N}{n}} = \frac{\binom{5}{t}\binom{95}{15-t}}{\binom{100}{15}}.$$

Thus

$$\begin{aligned} & P(X \leq 2) \\ = \; & P(X = 0) + P(X = 1) + P(X = 2) \\ = \; & \frac{\binom{5}{0}\binom{95}{15}}{\binom{100}{15}} + \frac{\binom{5}{1}\binom{95}{14}}{\binom{100}{15}} + \frac{\binom{5}{2}\binom{95}{13}}{\binom{100}{15}} \\ = \; & 0.4357 + 0.4034 + 0.1377 \\ = \; & 0.9768. \end{aligned}$$

(c) To find the O.C. for the sampling plan, repeat the above calculation for various values of p.

p	0.00	0.05	0.06	0.10	0.15	0.20	0.25	0.30
Pa	1.00	0.98	0.96	0.83	0.60	0.38	0.21	0.11

(d) If $p = 0.06$ then $Pa = 0.96$; if $p = 0.07$ then $Pa = 0.93$. Hence, for this sampling plan, the AQL lies in the interval $[0.06, 0.07]$.

(12) In this question we explore "double sampling plans". Obviously one could have "triple" or even more complicated sampling plans.

(a) The following tree diagram describes this sampling plan.

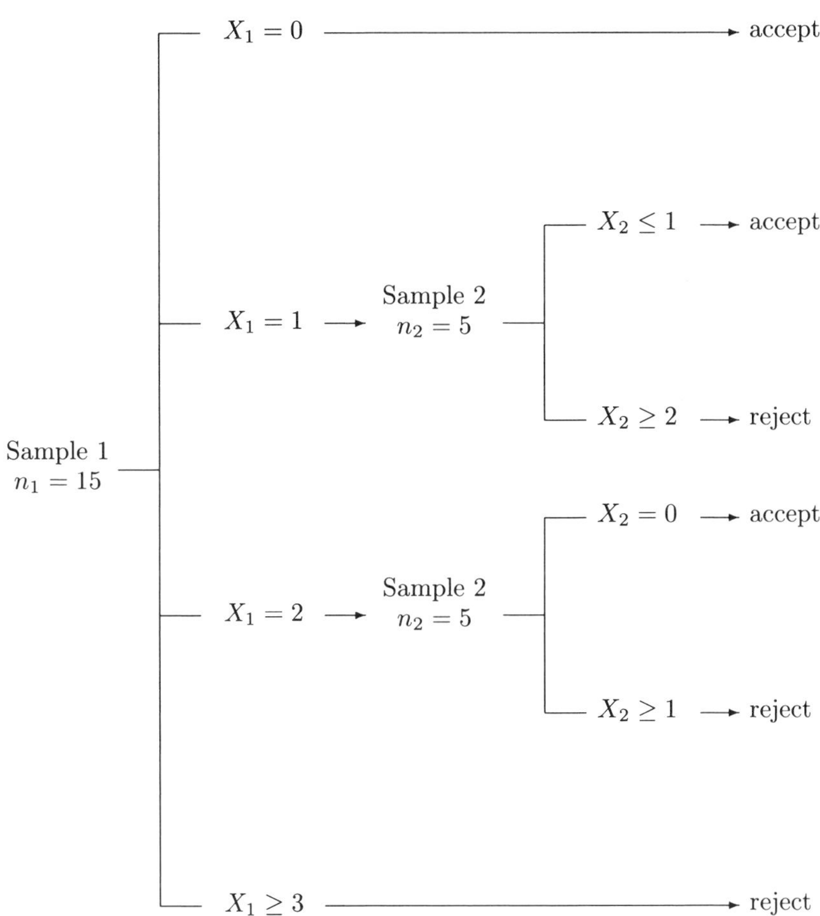
Sample 1
$n_1 = 15$
$X_1 = 0$
accept
$X_1 = 1$
Sample 2
$n_2 = 5$
$X_2 \leq 1$
accept
$X_2 \geq 2$
reject
$X_1 = 2$
Sample 2
$n_2 = 5$
$X_2 = 0$
accept
$X_2 \geq 1$
reject
$X_1 \geq 3$
reject

(b) The r.v.s X_1 and X_2 are independent of each other. The tree diagram allows us to write down the formula for Pa in terms of p.

$$\begin{aligned} & Pa \\ = \; & P(X_1 = 0) + P(X_1 = 1)P(X_2 \le 1) \\ & \quad + P(X_1 = 2)P(X_2 = 0) \\ = \; & (1-p)^{15} + 15p(1-p)^{14}\left((1-p)^5 + 5p(1-p)^4\right) \\ & \quad + \binom{15}{2} p^2(1-p)^{13}(1-p)^5 \\ = \; & (1-p)^{15} + 15p(1-p)^{19} + 180p^2(1-p)^{18}. \end{aligned}$$

(c) Points for the O.C. can be found by calculation.

p	0.00	0.04	0.05	0.10	0.15	0.20	0.25	0.30
Pa	1.00	0.96	0.93	0.68	0.41	0.21	0.09	0.04

(d) If $Pa = 0.95$ then p lies in the interval $[0.04, 0.05]$.

(e) Thus, the AQL for this double sampling plan lies in the interval $[0.04, 0.05]$ whereas the AQL for the single sampling plan lies in the interval $[0.06, 0.07]$. Comparing tables shows that poor quality batches (e.g. $p = 0.10$) have less chance of being accepted in the double sampling plan than in the single sample plan. This argument shows how O.C. functions can be used to assess different acceptance sampling schemes.

(13) This question examines the effect of linear transformations on rvs which are uniformly distributed.

(a) Let $Y = AX + B$. For $k = 0, 1, \ldots, N-1$,

$$P(Y = kA + B) = P(AX + B = kA + b) = P(X = k) = 1/N.$$

Thus Y is uniformly distributed over the set

$$\{B, A+B, 2A+B, \ldots, (N-1)A+B\}.$$

Thus, linear transformations of uniformly distributed rvs are uniformly distributed.

(b) Choose a number Z at random from the interval [0,1]; this can be done with a pocket calculator by using random number button. Let $X = 1 + [250Z]$. Then X is uniformly distributed over the set $\{0, 1, 2, \ldots, 249\}$ and hence $Y = 1+X$ is uniformly distributed over the set $\{1, 2, 3, \ldots, 250\}$.

(14) Perhaps you too could become as famous as Weldon by carrying out such an experiment!

(a) The rv X has a binomial distribution with parameters $n = 12$ and $p = 0.5$. Thus,

$$P(X = t) = \binom{12}{t} p^t (1-p)^{12-t} \quad (t = 0, 1, \ldots, 12).$$

(b) Weldon chose to conduct 4096 binomial experiments because $2^{12} = 4096$ and hence all the expected frequencies turn out to be whole numbers. In the table below we compare the frequencies observed by Weldon with the frequencies expected under the above binomial model. A χ^2 goodness-of-fit test leads to $\chi^2 = 34.586, \text{df} = 11, p < 0.001$. Thus, Weldon's data is not consistent with the model. A similar conclusion was drawn in [66, p. 338].

X	Binomial Probability	Expected Frequency	Observed Frequency
0	0.0002	1	0
1	0.0029	12	7
2	0.0161	66	60
3	0.0537	220	198
4	0.1208	495	430
5	0.1934	792	731
6	0.2256	924	948
7	0.1934	792	847
8	0.1208	495	536
9	0.0537	220	257
10	0.0161	66	71
11	0.0029	12	11
12	0.0002	1	0
Total	1.0000	4096	4096

Analysis of Weldon's data

(15) As said above, the Poisson distribution is the distribution of rare events. So one may expect that this would be the appropriate distribution to describe the frequency of strikes which commence in a week.

(a) $\overline{X} = 0.8994 \approx 0.90$.

(b) In the table below we compare the frequencies observed by with the frequencies expected under the above Poisson model. A χ^2 goodness-of-fit test leads to $\chi^2 = 0.7976, \text{df} = 3, p = 0.85$. Thus, the data is consistent with the model.

X	Poisson Probability	Expected Frequency	Observed Frequency
0	0.4066	254.5126	252
1	0.3659	229.0613	229
2	0.1647	103.0776	109
3	0.0494	30.9233	28
4	0.0111	6.9577	8
Total	0.9977	624.5326	626

Chapter 4

Continuous random variables

(1) For $p > 1$,

$$\int_1^\infty t^{-p} dt = 1/(p-1);$$

for $p \le 1$, this integral does not exist.

(a) Let

$$f_X(t) = \begin{cases} t^{-2} & t > 1 \\ 0 & t \le 1. \end{cases}$$

Then $f_X(t) \ge 0$ for all $t \in \Re$ and $\int_{-\infty}^\infty f_X(t)dt = 1$. Thus, f_X is a *bona fide* pdf. However

$$\int_{-\infty}^\infty t f_X(t)dt = \int_1^\infty t^{-1} dt$$

does not converge; hence μ_X does not exist.

(b) Let

$$f_X(t) = \begin{cases} 2t^{-3} & t > 1 \\ 0 & t \le 1. \end{cases}$$

Then

$$\mu_X = \int_{-\infty}^\infty t f_X(t)dt = 2\int_1^\infty t^{-2} dt = 2.$$

However

$$\int_{-\infty}^{\infty} t^2 f_X(t)dt = 2\int_1^{\infty} t^{-1}dt$$

does not converge; hence $E(X^2)$ does not exist; therefore σ_X^2 and σ_X do not exist.

(c) Let $p \geq 0$ and

$$f_X(t) = \begin{cases} (p+1)t^{-(p+2)} & t > 1 \\ 0 & t \leq 1. \end{cases}$$

Then, for $k \leq p$,

$$\begin{aligned} E(X^k) &= \int_{-\infty}^{\infty} t^k f_X(t)dt \\ &= (p+1)\int_1^{\infty} t^{k-(p+2)}dt = (p+1)/(p+1-k). \end{aligned}$$

However, for $k > p$,

$$\int_{-\infty}^{\infty} t^k f_X(t)dt = (p+1)\int_1^{\infty} t^{k-(p+2)}dt$$

does not converge; hence $E(X^k)$ does not exist.

(2) The gamma function $\Gamma : [0,\infty) \to [0,\infty)$ is defined by

$$\Gamma(\alpha) := \int_0^{\infty} \exp(-t)t^{\alpha-1}\,dt \quad (\alpha > 0). \tag{4.1}$$

The gamma function is one of the so-called "special functions" which are very important in applied mathematics. For an exposition on special functions by experts, see the recent book by Andrews, Askey and Roy [3].)

(a) When writing about the gamma function, many authors gloss over or ignore the technicality about whether $\Gamma(\alpha)$ is well defined by the above integral. The proof of this point can be found in the classic little book by E. Artin [5, pp. 11–12] or in books on mathematical analysis or advanced calculus such as Apostol [4, p.436]. Here we use Artin's proof.

To show that $\Gamma(\alpha)$ is well defined, we show that each of the integrals

$$I_1 = \int_0^1 \exp(-t)t^{\alpha-1}dt$$

and

$$I_2 = \int_1^\infty \exp(-t)t^{\alpha-1}dt$$

converges; then it will follow that the integral which defines $\Gamma(\alpha)$ converges.

For $0 < h < 1$,

$$0 < \int_h^1 \exp(-t)t^{\alpha-1}dt < \int_h^1 t^{\alpha-1}dt = \frac{1}{\alpha} - \frac{h^\alpha}{\alpha} < \frac{1}{\alpha}.$$

Thus, $\int_h^1 \exp(-t)t^{\alpha-1}dt$ is a bounded function of h which increases as $h \downarrow 0$. Thus,

$$\lim_{h\downarrow 0} \int_h^1 \exp(-t)t^{\alpha-1}dt$$

exists and hence I_1 converges.

Since $\exp(t) = 1 + t + t^2/2! + \ldots > t^n/n!$ for any $n > 0$, we have, for fixed $n > \alpha$,

$$\begin{aligned} 0 &< \int_1^x \exp(-t)t^{\alpha-1}dt \\ &< \int_1^x n!t^{\alpha-n-1}dt \\ &= \frac{n!}{n-\alpha}\left(1 - \frac{1}{x^{n-\alpha}}\right) \\ &< \frac{n!}{n-\alpha}. \end{aligned}$$

Thus $\int_1^x \exp(-t)t^{\alpha-1}dt$ is a bounded function of x which increases as $x \uparrow \infty$. Therefore I_2 converges.

(b) For $n \in \{1, 2, 3, \ldots\}$, let $P(n)$ denote the statement

$$P(n): \quad \Gamma(n) = (n-1)! \quad (n = 1, 2, 3, \ldots).$$

It is easy to see that $P(1)$ is true because

$$\Gamma(1) = \int_0^\infty \exp(-t)dt = 1 = 0!.$$

Assume that

$$P(n): \quad \Gamma(n) = (n-1)!$$

is true. Then

$$\begin{aligned}
\Gamma(n+1) &= \int_0^\infty t^n \exp(-t)dt \\
&= (-1)t^n \exp(-t) \mid_0^\infty + n\int_0^\infty \exp(-t)t^{n-1}dt \\
&= n\int_0^\infty \exp(-t)t^{n-1}dt \\
&= n\Gamma(n) \\
&= n(n-1)! \quad \text{(by inductive hypothesis)} \\
&= n!.
\end{aligned}$$

Thus, $P(n)$ is true for all $n \in \{1, 2, 3, \ldots\}$. This shows that the gamma function extends the factorial function.

(c) Let

$$f_X(t) = \begin{cases} \dfrac{1}{\Gamma(\alpha)\beta^\alpha} \exp\left(\dfrac{-t}{\beta}\right) t^{\alpha-1}, & (t > 0) \\ 0, & \text{(elsewhere)}. \end{cases}$$

Clearly $f_X(t) \geq 0$ for all $t \in \Re$. Furthermore,

$$\begin{aligned}
\int_\Re f_X(t)dt &= \frac{1}{\Gamma(\alpha)\beta^\alpha}\int_0^\infty \exp\left(\frac{-t}{\beta}\right) t^{\alpha-1}dt \\
&= \frac{1}{\Gamma(\alpha)}\int_0^\infty \exp(-u)u^{\alpha-1}du \quad (u = t/\beta) \\
&= 1 \quad \text{(by the definition of } \Gamma(\alpha)).
\end{aligned}$$

Thus, f_X is a well-defined pdf.
Let $k \in \{0, 1, 2, \ldots\}$. Then

$$\begin{aligned}
E(X^k) &= \frac{1}{\Gamma(\alpha)\beta^\alpha}\int_0^\infty \exp\left(\frac{-t}{\beta}\right) t^{k+\alpha-1}dt \\
&= \frac{\beta^k\Gamma(\alpha+k)}{\Gamma(\alpha)}.
\end{aligned}$$

(3) The Normal distribution is a source of many interesting calculations and patterns. Here $X \sim N(\mu, \sigma^2)$.

(a) Since

$$f_X(t) = \frac{1}{\sigma\sqrt{2\pi}} \exp -\frac{1}{2}\left(\frac{t-\mu}{\sigma}\right)^2 \quad (t \in \Re)$$

we have

$$f'_X(t) = \frac{(-1)(t-\mu)}{\sigma^3\sqrt{2\pi}} \exp -\frac{1}{2}\left(\frac{t-\mu}{\sigma}\right)^2 \quad (t \in \Re).$$

Hence $f'_X(t) = 0$ if and only if $t = \mu$. Thus f_X is unimodal and the mode of X is μ.
Observe that $f_X(\mu+t) = f_X(\mu-t)$ for all $t \in \Re$; therefore f_X is symmetric about the vertical axis $t = \mu$. Since

$$\int_{-\infty}^{\infty} f_X(t)dt = 1$$

it follows that

$$\int_{-\infty}^{\mu} f_X(t)dt = 0.5$$

and hence $F_X(\mu) = 0.5$. Thus μ is the median value of X.
Finally we calculate $E(X)$.

$$\begin{aligned}
\mu_X &= E(X) \\
&= \int_{-\infty}^{\infty} t f_X(t)dt \\
&= \frac{1}{\sigma\sqrt{2\pi}} \int_{-\infty}^{\infty} t \exp -\frac{1}{2}\left(\frac{t-\mu}{\sigma}\right)^2 dt \\
&= \frac{\sigma}{\sqrt{2\pi}} \int_{-\infty}^{\infty} z \exp(-z^2/2)dz + \frac{\mu}{\sqrt{2\pi}} \int_{-\infty}^{\infty} \exp(-z^2/2)dz \\
&\qquad (z = (t-\mu)/\sigma) \\
&= 0 + \mu = \mu.
\end{aligned}$$

Thus, the mean of X is μ.

(b) A little algebra leads to

$$f_X^{(2)}(t) = \frac{1}{\sigma^3\sqrt{2\pi}}\left(-1+\left(\frac{t-\mu}{\sigma}\right)^2\right) \exp -\frac{1}{2}\left(\frac{t-\mu}{\sigma}\right)^2.$$

Thus,

$$f_X^{(2)}(t) = 0 \Leftrightarrow (t-\mu)^2 = \sigma^2 \Leftrightarrow t = \mu \pm \sigma.$$

Furthermore, we note that $f_X^{(2)}(t)$ changes sign at each of $t = \mu \pm \sigma$. Thus the points of inflexion of f_X are $t = \mu \pm \sigma$.

(c) To prove that $E((X-\mu)^4) = 3\sigma^4$, we need the following results about the gamma function. For $\alpha > 0$, $\beta > 0$,

$$\begin{aligned} \Gamma(\alpha) &:= \int_0^\infty t^{\alpha-1}\exp(-t)dt; \\ \int_0^\infty t^{\alpha-1}\exp(-t/\beta)dt &= \Gamma(\alpha)\beta^\alpha; \\ \Gamma(\alpha+1) &= \alpha\Gamma(\alpha); \\ \Gamma(0.5) &= \sqrt{\pi}. \end{aligned}$$

These results can be found in standard reference books on special functions or the gamma function (e.g. Whittaker and Watson [71, p. 243 *et seq.*] one of the classic works in mathematical analysis). Thus

$$\begin{aligned} E((X-\mu)^4) &= \frac{1}{\sigma\sqrt{2\pi}}\int_{-\infty}^{\infty}(t-\mu)^4\exp -\frac{1}{2}\left(\frac{t-\mu}{\sigma}\right)^2 dt \\ &= \frac{2\sigma^4}{\sqrt{2\pi}}\int_0^\infty z^4\exp(-z^2/2)dx \\ &\quad \text{(where } z = (t-\mu)/\sigma\text{)} \\ &= \frac{\sigma^4}{\sqrt{2\pi}}\int_0^\infty u^{3/2}\exp(-u/2)du \\ &\quad \text{(where } u = z^2\text{)} \\ &= \frac{\sigma^4}{\sqrt{2\pi}}\Gamma(5/2)2^{(5/2)} \\ &= 3\sigma^4. \end{aligned}$$

(d) If you examine the definitions of the Normal distribution in some elementary textbooks on statistics, then you will find that the definitions are often vague, incomplete or inadequate. You are encountering the problem faced by teachers of statistics: how does one explain the Normal curve defined by the function

$$y = \frac{1}{\sigma\sqrt{2\pi}} \exp -\frac{1}{2}\left(\frac{t-\mu}{\sigma}\right)^2 \quad (t \in \Re)$$

to students who do not know about exponential functions? In an excellent elementary text Freund and Perles [26, p. 234] write

> "...the normal distribution is sometimes described simply as a bell-shaped distribution. This description leaves a good deal to be desired: it is true that normal distribution curves, often called normal curves, are bell-shaped, but not all bell-shaped distribution curves are necessarily those of normal distributions."

One way around this problem is to state simply that the Normal curve is approximately defined by the function

$$y = \frac{1}{\sigma\sqrt{2\pi}}(2.718)^{-\frac{1}{2}\left(\frac{t-\mu}{\sigma}\right)^2} \quad (t \in \Re).$$

Students with no calculus backgound can sketch this graph using just a calculator.

(e) For $x > 0$,

$$\begin{aligned}
\Phi(x) &= \frac{1}{\sqrt{2\pi}}\int_{-\infty}^{x} \exp(-t^2/2)dt \\
&= 0.5 + \frac{1}{\sqrt{2\pi}}\int_{0}^{x} \exp(-t^2/2)dt \\
&= 0.5 + \frac{1}{\sqrt{\pi}}\int_{0}^{x/\sqrt{2}} \exp(-u^2)dt \quad (u = t/\sqrt{2}) \\
&= 0.5 + \frac{1}{2}\text{erf}(x/\sqrt{2}).
\end{aligned}$$

So

$$2\Phi(x) = 1 + \text{erf}(x/\sqrt{2}).$$

(f) This result is the key to using the standard Normal distribution tables to obtain probabilites associated with Normal distributions other than the standard Normal distribution. Let $Z = (X - \mu)/\sigma$.

$$\begin{aligned} F_Z(t) &= P(Z < t) \\ &= P(Z = (X - \mu)/\sigma < t) \\ &= P(X < \mu + \sigma t) \\ &= F_X(\mu + \sigma t). \end{aligned}$$

So

$$\begin{aligned} f_Z(t) &= F'_Z(t) \\ &= \sigma F'_X(\mu + \sigma t) \\ &= \sigma f_X(\mu + \sigma t) \\ &= \frac{1}{\sqrt{2\pi}} \exp(-t^2/2). \end{aligned}$$

Thus $Z \sim N(0, 1)$.

(4) One way of developing your skills in mathematical modelling is to look continually at the way in which mathematical ideas can be related to the world around us. The point of this exercise is to encourage you to see probability distributions in various situations.

(a) We present two examples.

Example 1: Most pocket calculators have a random number generator button. This facility produces the value of a r.v. which is uniformly distributed over the interval $(0, 1)$. Generating such values is important in many simulation experiments which are based on the uniform distribution. For example, Box and Muller [9] prove that if

- U_1 and U_2 are independent and uniformly distributed over $(0, 1)$
- $Z_1 = \sqrt{-2 \ln U_1} \cos(2\pi U_2)$
- $Z_2 = \sqrt{-2 \ln U_1} \sin(2\pi U_2)$

then Z_1 and Z_2 are independent and $N(0, 1)$. Thus the uniform distribution is the basis of generating values of standard Normal r.vs. and hence any Normal r.v.

Example 2: Suppose that a rather inexperienced player is throwing darts at a dart-board. We give polar coordinates

(r, θ) to each point on the board. If we record θ for each throw then it is a reasonable hypothesis that θ is uniformly distributed over the interval $(0, 2\pi)$.

(b) Many biological measurements are assumed to be Normally distributed. Here are two examples.

Example 1: Let X be the weight of a fully grown male Chihuahua chosen at random from the population in Australia. (Notice how carefully one defines the random variable. We define a population precisely; we state that we are choosing an individual at random from this population — this makes our variable "random"; and finally we define exactly what we are measuring to ensure that we are observing a "variable".) It is reasonable to suggest that X is Normally distributed. From leafing through various books on this breed of dog, I suggest that μ_X is 2 kg. I would also guess that for almost all such dogs, $1.5 \leq X \leq 2.5$. Now almost all values of a Normally distributed r.v. lie in $\mu_X - 3\sigma_X \leq X \leq \mu_X + 3\sigma_X$ and hence I would guess that $6\sigma_X$ is about 1.00. Hence a sensible hypothesis is that

$$X \sim N(2.00, 0.17^2).$$

Example 2: Many statistical methods rest on the assumption that the r.v. under consideration is Normally distributed. However, often times, this assumption is made without being thoroughly tested. For example, manufacturers of mohair may wish to apply various quality control measures to the diameters of goat hairs. Let X be the diameter of a goat hair chosen at random from a particular breed of goat. It is reasonable to suggest that this r.v. is Normally distributed; to estimate the mean and variance one would go through the argument as in the previous example. The problem is that to test such a hypothesis, one needs a lot of data. It is worthwhile for someone to take the trouble to examine this hypothesis carefully, with a lot of data, and settle the matter. Then the industry has a substantial basis for adopting the assumption that such measurements are Normally distributed when such an assumption is necessary for a particular quality assurance procedure.

(c) Let X be the age of a student chosen at random from the population of students at La Trobe University, Bendigo; age will be measured on a continuous scale in years. So, for a student who is 21 years and 2 months, we will write $X = 21.167$. I imagine that the distribution of X would be skewed for the following reasons.

- There would be almost no students under the age of 17.0 years.
- Many students would have $19.00 \leq X \leq 20$ because a very large percentage of first year students will be in this group.
- There will be another large group of older students with $25 \leq X \leq 40$.
- We will find students in the range $50 \leq X \leq 70$ or even older.

Thus, the distribution will not be symmetric.

(5) Again we encounter the difficulties associated with introducing mathematical ideas associated with statistics to students who want to know about statistics but do not have the time or interest to study the associated ideas in mathematics.

(a) In [26, p. 75] the empirical rule is stated as follows.

> For distributions having the general shape of the cross section of a bell,
>
> (1) about 68% of the values will lie within 1 standard deviation of the mean, that is between $\overline{x} - s$ and $\overline{x} + s$;
> (2) about 95% of the values will lie within 2 standard deviations of the mean, that is between $\overline{x} - 2s$ and $\overline{x} + 2s$;
> (3) about 99.7% of the values will lie within 3 standard deviations of the mean, that is between $\overline{x} - 3s$ and $\overline{x} + 3s$.

> This result is sometimes referred to as the empirical rule, although actually it is a theoretical result based on the normal distribution...

The book is an excellent introductory book on statistics and the authors do, in the last sentence, hint that this rule is not applicable to any bell-shaped distribution. Many elementary texts give no such hint.

(b) Let $Z \sim N(0,1)$. Then from a set of tables of the standard Normal distribution we find:

$$\begin{aligned} P(|Z| \leq 1) &= 0.6826, \\ P(|Z| \leq 2) &= 0.9544, \\ P(|Z| \leq 1) &= 0.9974. \end{aligned}$$

Thus, the "empirical rule" does indeed work for a standard Normal distribution.

(c) Here we will see that although the t-distribution curve with $\nu = 3$ d.f., is a nice bell-shaped curve, the empirical rule does not work so well.

$$f_X(t) = \frac{2}{\pi} \cdot \frac{1}{(1+t^2)^2} \quad (t \in \Re).$$

i. The function f_X is an even function and hence symmetric about $t = 0$. We have, for $t \in \Re$,

$$\begin{aligned} f'_X(t) &= -\frac{8}{\pi} \cdot \frac{t}{(1+t^2)^3}; \\ f^{(2)}_X(t) &= \frac{8}{\pi} \cdot \frac{5t^2 - 1}{(1+t^2)^4}. \end{aligned}$$

Thus, $f'_X(t) = 0$ if and only if $t = 0$ and $f^{(2)}_X(t) = 0$ if and only if $t = \pm 1/\sqrt{5}$. From careful examination of the first two derivatives of f_X we see that the graph of f_X is indeed a bell-shaped curve which is symmetric about $t = 0$.

ii. We evaluate $E(X) = \mu_X$ and σ_X in the usual way.

$$\begin{aligned} E(X) &= \int_{\Re} \frac{2}{\pi} \cdot \frac{t}{(1+t^2)^2} dt \\ &= 0; \end{aligned}$$

$$\begin{aligned}
E(X^2) &= \int_{\Re} \frac{2}{\pi} \cdot \frac{t^2}{(1+t^2)^2} dt \\
&= \frac{4}{\pi} \int_0^{\infty} \frac{t^2}{(1+t^2)^2} dt \\
&= \frac{4}{\pi} \int_0^{\infty} \frac{1}{(1+t^2)^2} - \frac{1}{(1+t^2)^2} dt \\
&= \frac{4}{\pi} \int_0^{\pi/2} (1-\cos^2\theta) d\theta \quad (t = \tan\theta) \\
&= 1; \\
\sigma_X^2 &= E(X^2) - E(X)^2 \\
&= 1.
\end{aligned}$$

iii. You can use a package such as `Excel` to show that if $X \sim t(3)$ then

$$\begin{aligned}
P(|X-\mu_X| \leq \sigma_X) &= P(|X| \leq 1) \\
&\approx 0.61; \\
P(|X-\mu_X| \leq 2\sigma_X) &= P(|X| \leq 2) \\
&\approx 0.86.
\end{aligned}$$

(d) This example illustrates that the "empirical rule" is meant to apply to a Normal distribution, not just any bell-shaped distribution.

(6) In an excellent recent book, Jacod and Protter [35, p. 39] suggest that the Cauchy distribution was created as a source of counterexamples. Here we see why!
Let $X \sim$ Cauchy.

(a) The graph of the function f_X is symmetric about $x = 0$ and "bell-shaped". However, we will see below that the properties of this distribution are very different from those of the Normal distibution. Thus the phrase "bell-shaped" does not identify the Normal distribution despite suggestions to the contrary in many text books.

(b) The fact that $E(X)$ does not exist follows from the fact that

$$\int_{\Re} t f_X(t) dt = \int_{\Re} \frac{t}{\pi(1+t^2)} dt$$

does not converge. Recall that $\int_{\Re} \ldots dt$ exists if and only if each of

$$\int_{-\infty}^{0} \ldots dt \quad \text{and} \quad \int_{0}^{\infty} \ldots dt$$

exist. However

$$\begin{aligned} & \int_0^\infty \frac{t}{\pi(1+t^2)} dt \\ = \ & \lim_{x\to\infty} \int_0^x \frac{t}{\pi(1+t^2)} dt \\ = \ & \lim_{x\to\infty} \ln(1+x^2). \end{aligned}$$

Thus $\int_0^\infty \dfrac{t}{\pi(1+t^2)}$ does not exist and hence $E(X)$ does not exist.

(c) Let $Y = 1/X$. Then

$$\begin{aligned} F_Y(t) &= P(Y \le t) \\ &= P(X \ge 1/t) \\ &= \frac{1}{\pi} \int_{1/t}^{\infty} \frac{1}{1+u^2} du \\ &= \frac{1}{\pi} \tan^{-1} u|_{1/t}^{\infty} \\ &= \frac{1}{\pi} \left(\frac{\pi}{2} - \tan^{-1}(1/t) \right); \\ f_Y(t) &= F_Y'(t) \\ &= \frac{1}{\pi(1+t^2)} \end{aligned}$$

— which is the pdf of the Cauchy distribution!

(d) Consider the random variable Y with p.d.f.

$$f_Y(t) = \begin{cases} 1/4 & (|t| \le 1) \\ 1/(4x^2) & (|t| > 1) \end{cases}$$

Let $W = 1/Y$. For $t > 1$, $0 < 1/t < 1$ and

$$\begin{aligned} F_W(t) &= P(W \le t) \\ &= P(Y \ge 1/t) \\ &= 1 - P(Y < 1/t) \end{aligned}$$

$$
\begin{aligned}
&= 1 - \int_{-\infty}^{1/t} f_Y(u)du \\
&= 1 - \int_{-\infty}^{0} f_Y(u)du - \int_{0}^{1/t} f_Y(u)du \\
&= 0.5 - \frac{1}{4t}; \\
f_W(t) &= F'_W(t) \\
&= \frac{1}{4t^2} \\
&= f_Y(t).
\end{aligned}
$$

Similar arguments can be applied for other values of t. Thus $1/Y$ has the same distribution as Y.

(7) The proofs for the continuous case can be proved easily by mimicking the proofs for the discrete case. Just follow your nose and replace $\sum$ by $\int$ where appropriate. You may wonder if there is some all encompassing approach so that we can "kill two birds with the one stone". Once you ask this sort of question, you are ready to tackle probability theory via measure theory. There are many excellent texts on measure theory and probability. Rosenthal [56] is new, short and accessible, so too is Jacod and Protter [35], Kolmogorov [40] is where it all started, — but my favourite is Shiryaev [60].
We illustrate with a proof of Chebyshev's inequality for the continuous case.

$$
\begin{aligned}
E(X^2) &= \int_{\Re} t^2 f_X(t) \\
&= \int_{|t|<c} t^2 f_X(t)\, dt + \int_{|t|\geq c} t^2 f_X(t)\, dt \\
&\geq c^2 \int_{|t|\geq c} f_X(t)\, dt \\
&= c^2 P(|X| \geq c); \\
P(|X| \geq c) &\leq c^{-2} E(X^2).
\end{aligned}
$$

Other parts are proved in a similar fashion.

(8) Vilfredo Pareto (1848–1923) was an Italian economist and sociologist although he was initially trained in mathematics and engineering. He applied his skills in mathematics to economics, especially to understanding the distribution of income — hence we have the Pareto distribution [36, p. 575].

We say that the r.v. X has a Pareto distribution with parameter 4 if

$$f_X(t) = \begin{cases} kt^{-4} & (t > 1) \\ 0 & (t \leq 1). \end{cases}$$

where k is some positive constant.

(a) To find k, note that

$$\begin{aligned} 1 &= \int_{-\infty}^{\infty} f_X(t)dt \\ &= k\int_1^{\infty} t^{-4}dt \\ &= k/3. \end{aligned}$$

Thus, $k = 3$.

Note the advantage of assuming that the r.v. is always distributed over the entire real line. This way we avoid worrying about defining the end points of the interval of integration.

(b) The d.f. F_X is given by:

$$\begin{aligned} F_X(t) &= P(X \leq t) \\ &= \int_{-\infty}^{t} f_X(u)du \\ &= \begin{cases} 3\int_1^t u^{-4}du, & t > 1 \\ 0, & t \leq 1. \end{cases} \\ &= \begin{cases} 1 - (1/t)^3, & t > 1 \\ 0, & t \leq 1. \end{cases} \end{aligned}$$

(c) The mean μ_X and s.d. σ_X are given by:

$$\begin{aligned} \mu_X &= \int_{-\infty}^{\infty} tf_X(t)dt \\ &= 3\int_1^{\infty} t^{-3}dt \end{aligned}$$

$$\begin{aligned}
&= 3/2; \\
E(X^2) &= \int_{-\infty}^{\infty} t^2 f_X(t)dt \\
&= 3\int_1^{\infty} t^{-2}dt \\
&= 3; \\
\sigma_X^2 &= E(X^2) - E(X)^2 \\
&= 3/4; \\
\sigma_X &= \sqrt{3}/2.
\end{aligned}$$

(d) Therefore

$$\begin{aligned}
& P(|X - \mu_X| \leq 1.5\sigma_X) \\
= \;& P(|X - 3/2| \leq 3\sqrt{3}/4) \\
= \;& P\left(\frac{6 - 3\sqrt{3}}{4} < X < \frac{6 + 3\sqrt{3}}{4}\right) \\
= \;& P(1 < X < \frac{6 + 3\sqrt{3}}{4}) \\
& \text{(because } \frac{6 - 3\sqrt{3}}{4} < 1) \\
= \;& 3\int_1^{\frac{6+3\sqrt{3}}{4}} t^{-4}dt \\
\approx \;& 0.954.
\end{aligned}$$

(e) According to Chebyshev's inequality

$$P(|X - \mu_X| \leq 1.5\sigma_X) \geq 1 - \left(\frac{1}{1.5}\right)^2 \approx 0.556.$$

Chebyshev's estimate is consistent with the exact value found in the previous part. You may think that Chebyshev's estimate is woefully inaccurate; on the other hand, Chebyshev assumes knowledge of only μ_X and σ_X; to me it is amazing that you can say **anything** with so little knowledge!

(9) Recall the information concerning the Gamma distribution in §4.1.

(a) If X has an exponential distribution with parameter λ then the p.d.f. of X is given by

$$f_X(t) = \begin{cases} \lambda \exp(-\lambda t) & (t > 0) \\ 0 & (t \le 0). \end{cases}$$

Thus the d.f. of X is given by

$$\begin{aligned} F_X(t) &= \int_{-\infty}^{t} f_X(t) \\ &= \begin{cases} \int_0^t \lambda \exp(-\lambda u)\, dt & (t > 0) \\ 0 & (t \le 0) \end{cases} \\ &= \begin{cases} 1 - \exp(-\lambda t) & (t > 0) \\ 0 & (t \le 0) \end{cases} \end{aligned}$$

(b) The graphs of f_X and F_X can be easily sketched even by plotting points!

(c) Let $Y = F_X(X)$; then $0 \le Y \le 1$. Let $t \in [0, 1]$. Then

$$\begin{aligned} F_Y(t) &= P(Y \le t) \\ &= P(F_X(X) \le t) \\ &= P(1 - \exp(-\lambda X) \le t) \\ &= P(\exp(-\lambda X) \ge 1 - t) \\ &= P(-\lambda X \ge \ln(1 - t)) \\ &= P(X \le \frac{-1}{\lambda} \ln(1 - t)) \\ &= 1 - \exp(\ln(1 - t)) \quad \text{by above expression for } F_X \\ &= 1 - (1 - t) \\ &= t. \end{aligned}$$

Thus,

$$f_Y(t) = F_Y'(t) = 1 \quad (0 \le t \le 1)$$

and hence, $Y \sim \text{Uniform}(0, 1)$.

(d) This problem illustrates a basic approach to generating random samples from a given distribution. Although it works well for the exponential distribution, this approach is not always effective.

Use a pocket calculator or a table of random numbers to generate 5 observations of Y where $Y \sim \text{Uniform}(0, 1)$. My calculator gave:

$$Y_1 = 0.645, \quad Y_2 = 0.479, \quad Y_3 = 0.210,$$

$$Y_4 = 0.414, \quad Y_5 = 0.902.$$

We saw above that

$$\begin{aligned} Y &= F_X(X) \\ &= 1 - \exp(-\lambda X). \end{aligned}$$

So, since $\lambda = 2$,

$$X = \frac{-1}{2} \ln(1 - Y).$$

Being able to write down, explicitly, the function F_X^{-1} is the key property of the exponential distribution which makes this basic approach feasible in this particular case. This leads to

$$X_1 = 0.5178, \quad X_2 = 0.3260, \quad X_3 = 0.1179,$$

$$X_4 = 0.2672, \quad X_5 = 1.1614$$

which is a random sample of $n = 5$ observations from an exponential distribution with parameter $\lambda = 2$.

(10) Another application of the log normal distribution in health care recently came to my attention. An important quality indicator in health care is length of stay in hospital for patients with a particular condition. It is often assumed that the logarithm of length of stay is Normally distributed; that it, length of stay has a log normal distribution.

(a) We say that the distribution of Y is "log normal" because its logarithm has a Normal distribution. In this exercise, $\ln(Y) \sim N(0, 1)$; however one could consider, more generally, a r.v. Y such that $\ln(Y) \sim N(\mu, \sigma^2)$.

(b) We find f_Y as follows.

$$\begin{aligned} F_Y(t) &= P(Y \leq t) \\ &= P(\exp(X) \leq t) \\ &= P(X \leq \ln(t)) \\ &= \Phi(\ln(t)). \end{aligned}$$

Thus

$$\begin{aligned} f_Y(t) &= F_Y'(t) \\ &= \phi(\ln t).\frac{1}{t} \\ &= \frac{1}{\sqrt{2\pi}}\frac{1}{t}\exp(-\frac{(\ln t)^2}{2}). \end{aligned}$$

(c) Sketch the graph of f_Y.

(11) Let $X \sim N(0,1)$. To calculate $P(0 \leq X \leq 1)$ we need to calculate the integral

$$I = P(0 \leq X \leq 1) = \frac{1}{\sqrt{2\pi}}\int_0^1 \exp(-t^2/2)\, dt.$$

This rather harmless looking intergal is impossible to calculate exactly using the standard techniques in systematic integration. Thus, we resort to some numerical technique such as the trapezoidal rule or Simpson's rule; these methods are described in standard calculus books such as Stewart [63, pp. 546–556]. The problem of calculating integrals of the function $\exp(-t^2/2)$ is used to demonstrate a great variety of numerical methods throughout the text by Kahaner, Moler and Nash [37].

Here we apply Simpson's rule. Divide $[0,1]$ into an even number, (say $2n = 10$) of subintervals of equal size $h = 1/(2n) = 0.1$:

$$0 = t_0 < t_1 < t_2 < \ldots < t_9 < t_{10}.$$

Simpson's rule tells us that I is approximately

$$\phi(t_0)+4\phi(t_1)+2\phi(t_2)+4\phi(t_3)+2\phi(t_4)+\ldots+2\phi(t_8)+4\phi(t_9)+\phi(t_{10})$$

where

$$\phi(t) = \frac{1}{\sqrt{2\pi}}\exp(-t^2/2).$$

This leads to $I \approx 0.341345016$.
There are many papers which discuss methods for approximating the Normal distribution function Φ: see Patel and Read [50, Chapter 3].
For example, Moran [46] presents the approximation

$$\Phi(t) \approx 0.5+\frac{1}{\pi}\sum_{n=0}^{12}(n+0.5)^{-1}\exp(-(n+0.5)^2/9)\sin\left(\frac{(n+0.5)t\sqrt{2}}{3}\right)$$

and this leads to $I = \Phi(1) - 0.5 \approx 0.341344746$; Moran states that his approximation is correct to 9 decimal places for $|t| \leq 7$. Algorithms like these cited are built into packages such as `Excel` which, by the way, gives $I \approx 0.34134474$.

(12) This result is an early result in the beautiful subject known as geometric probability: for a short introduction see Kendall and Moran [38]. We have $L < 1$ to ensure that the needle will intersect at most one horizontal line.

(a) Assume that $y \sim \text{Uniform}(0, 0.5)$ and $\theta \sim \text{Uniform}(0, \pi/2)$ are independent r.v. From the geometry of the situation, it is clear that the needle intersects the horizontal line if

$$y \leq \frac{L}{2}\cos\theta.$$

So we must calculate the probability of such an event when y and θ have the above distributions. This is easily done if the problem is set in the $\theta - y$ plane with $0 \leq \theta \leq \pi/2$ on the horizontal axis and $0 \leq y \leq 0.5$ on the vertical axis. Let $A = (0,0)$, $B = (\pi/2, 0)$, $C = (\pi/2, 0.5)$, $D = (0, 0.5)$. Draw the graph of

$$y \leq \frac{L}{2}\cos\theta \quad (0 \leq \theta \leq \pi/2)$$

remembering that $0 < L < 1$. Then the required probability is

$$\frac{\int_0^{\pi/2}(L/2)\cos\theta\, d\theta}{\text{Area }(ABCD)} = \frac{2L}{\pi}.$$

(b) An experiment for estimating π can be devised as follows. Throw such a needle N times where N is large; let X denote the number of throws which result in the needle crossing a line; then

$$\frac{X}{N} \approx \frac{2L}{\pi}$$

and hence

$$\pi \approx \frac{2L}{X}.$$

Many such tedious experiments have been reported in the literature; see [38, p. 70] — where the authors say:

> "...a better method for estimating π is to cut out a large circle of wood and use a tape measure."

However, Buffon's problem does give an amusing introduction to the beautiful but difficult subject of geometric probability. On a historical note, George Louis Leclerc Buffon (1707–1788) was a distinguished French naturalist. According to Barry Jones [36] in some ways, Buffon's ideas anticipated Darwin's theory of evolution. For more information on Buffon, and many other statisticians, I refer the reader to new collection of essays by C. Heyde and E. Seneta [34] Indeed, if you are interested in historical aspects of probability and statistics, this collection by two distinguished scholars is an excellent starting point.

(13) If

$$p(x) = x^2 + Bx + C$$

then p has real roots if and only if $B^2 > 4C$. We may regard (B, C) as a point chosen at random in the square $[0,1] \times [0,1]$. Since the area of this square is 1, we can say that

$$\begin{aligned}
& P(p \text{ has real roots}) \\
= \; & P(B^2 > 4C) \\
= \; & P((B,C) \text{ lies under the curve } C = B^2/4, \ (0 < B < 1)) \\
= \; & 0.25 \int_0^1 B^2 dB \\
= \; & 1/12.
\end{aligned}$$

In the theory of random polynomials, we have the opportunity to explore well-known facts about ordinary polynomials with non-random coefficients in a more general setting. For example, we know that a polynomial of degree n can have at most n real zeros.

- What is the mean value of the number of real zeros of a random polynomial of degree n whose coefficients have a given distribution?
- What is the variance of the number of real zeros of a random polynomial of degree n whose coefficients have a given distribution?

For answers to these, and similar fascinating questions, you can refer to the standard reference in the theory of random polynomials by Bharucha-Reid and Sambandham [8].

Chapter 5

Limit theorems

(1) This is a long question which takes you through the steps of proving this important result. Although the result is important in probability theory, the proof is often omitted from probability texts. I hope that you enjoy the journey which is typical of the reasoning used in proving similar results in classical analysis.

(a) This is straightforward.

$$\begin{aligned} S(n) &= \log(n!) \\ &= \log \prod_{k=1}^{n} k \\ &= \log \prod_{k=2}^{n} k \\ &= \log \prod_{p=1}^{n-1} (p+1) \\ &= \sum_{p=1}^{n-1} \log(p+1). \end{aligned}$$

(b) Using the definitions of $A(p)$, $B(p)$ and $C(p)$ we find

$$S(n) = \sum_{p=1}^{n-1} \log(p+1)$$

$$
\begin{aligned}
&= \sum_{p=1}^{n-1} A(p) + B(p) - C(p) \\
&= \sum_{p=1}^{n-1} A(p) + \sum_{p=1}^{n-1} B(p) - \sum_{p=1}^{n-1} C(p) \\
&= \sum_{p=1}^{n-1} \int_p^{p+1} \log t \, dt + \sum_{p=1}^{n-1} \left(\log(p+1) - \log p\right)/2 \\
&\quad - \sum_{p=1}^{n-1} C(p) \\
&= \int_1^n \log t \, dt + 0.5 \log n - \sum_{p=1}^{n-1} C(p) \\
&= (t \log t - t)|_1^n + 0.5 \log n - \sum_{p=1}^{n-1} C(p) \\
&= (n + 0.5) \log n - n + 1 - \sum_{p=1}^{n-1} C(p).
\end{aligned}
$$

(c) Using the definition of $C(p)$,

$$
\begin{aligned}
C(p) &= \int_p^{p+1} \log t \, dt - \left(\log(p+1) - \log p\right)/2 \\
&= (t \log t - t)|_p^{p+1} - \left(\log(p+1) - \log p\right)/2 \\
&= (p + 1/2) \log \left(\frac{p+1}{p}\right) - 1.
\end{aligned}
$$

(d) In [18, p. 319] we find

$$
\log \left(\frac{1+x}{1-x}\right) = 2 \sum_{k=1}^{\infty} \frac{x^{2k-1}}{2k-1} \quad (|x| < 1).
$$

This means that the right hand side converges (pointwise) for each value of $x \in (-1, 1)$. Furthermore, if we choose $0 < \rho < 1$ then it can be shown that the right hand side converges to the left hand side uniformly in $[-\rho, \rho]$.

Notice that we do not bother to prove this fact here. It is sufficient for our purposes to check (not merely find) the result in a reputable source. This is part of the mathematical research process.

(e) If

$$\frac{p+1}{p} = \frac{1+x}{1-x}$$

then

$$x = \frac{1}{1+2p}.$$

So

$$\begin{aligned} C(p) &= (p+1/2)\log\left(\frac{p+1}{p}\right) - 1 \\ &= 2(p+1/2)\sum_{k=1}^{\infty}\frac{1}{(2k-1)(1+2p)^{2k-1}} - 1 \\ &= (2p+1)\sum_{k=2}^{\infty}\frac{1}{(2k-1)(1+2p)^{2k-1}} \\ &= \sum_{k=1}^{\infty}\frac{1}{(2k+1)(2p+1)^{2k}}. \end{aligned}$$

(f) For $k \geq 1$, we have $2k+1 \geq 3$. Thus,

$$\begin{aligned} C(p) &\leq \frac{1}{3}\sum_{k=1}^{\infty}\frac{1}{(2p+1)^{k}} \\ &= \frac{1}{3}\frac{1}{(2p+1)^2-1} \\ &= \frac{1}{12p(p+1)} \\ &= \frac{1}{12}\cdot\left(\frac{1}{p} - \frac{1}{p+1}\right). \end{aligned}$$

(g) The calculations in this part do not seem to be as "natural" as those in the previous part; they appear to be forced or contrived. Indeed, they are contrived in order to make the lower bound for $C(p)$ resemble as closely as possible the upper bound just obtained.

$$\begin{aligned}
C(p) &= \sum_{k=1}^{\infty} \frac{1}{(2k+1)(2p+1)^{2k}} \\
&= \frac{1}{3(2p+1)^2} + \frac{1}{5(2p+1)^4} + \frac{1}{7(2p+1)^6} + \dots \\
&> \frac{1}{3(2p+1)^2} + \frac{1}{3^2(2p+1)^4} + \frac{1}{3^3(2p+1)^6} + \dots \\
&= \frac{1}{3(2p+1)^2} \sum_{k=0}^{\infty} 3^{-k}(2p+1)^{-2k} \\
&= \frac{1}{12(p^2+p+(1/6))}
\end{aligned}$$

So far, our calculations have been straightforward. We now come to the delicate part. It would be very helpful if we could write

$$\frac{1}{p^2+p+(1/6)} \geq \frac{1}{(p+a)(p+1+a)}$$

because then we would have

$$C(p) > \frac{1}{12(p^2+p+(1/6))} \geq \frac{1}{12}\left(\frac{1}{p+a} - \frac{1}{p+1+a}\right)$$

and harking back to the upper bound for $C(p)$ we can see the utility of such an expression. So we need to find a such that

$$p^2 + p + (1/6) \leq (p+a)(p+1+a)$$

and a little checking shows that $a = 1/12$ will do the job. (In fact, $a = 1/18.32$ would do the job but it makes little difference.) Thus,

$$C(p) \quad > \quad \frac{1}{12} \cdot \left(\frac{1}{p+(1/12)} - \frac{1}{p+1+(1/12)}\right).$$

(h) If $D = \sum_{p=1}^{\infty} C(p)$ then

$$D \quad < \quad \sum_{p=1}^{\infty} \frac{1}{12}\left(\frac{1}{p} - \frac{1}{p+1}\right)$$

$$\begin{aligned} &= \frac{1}{12}\left(\left(1-\frac{1}{2}\right)+\left(\frac{1}{2}-\frac{1}{3}\right)+\left(\frac{1}{3}-\frac{1}{4}\right)+\dots\right) \\ &= \frac{1}{12} \end{aligned}$$

and

$$\begin{aligned} D &> \sum_{p=1}^{\infty}\frac{1}{12}\left(\frac{1}{p+(1/12)}-\frac{1}{p+1+(1/12)}\right) \\ &= \frac{1}{12}\Big[\left(\frac{1}{1+(1/12)}-\frac{1}{2+(1/12)}\right) \\ &\quad +\left(\frac{1}{2+(1/12)}-\frac{1}{3+(1/12)}\right) \\ &\quad +\left(\frac{1}{3+(1/12)}-\frac{1}{4+(1/12)}\right)+\dots\Big] \\ &= \frac{1}{13}. \end{aligned}$$

(If we had used $a = 1/18.32$ we would have ended up with $D > 1/12.7$.) Hence

$$\frac{1}{13} < D < \frac{1}{12}$$

and similarly,

$$\frac{1}{12n+1} < r_n < \frac{1}{12n}.$$

(i) We have

$$\begin{aligned} S_n &= (n+0.5)\ln n - n + 1 - \sum_{p=1}^{n-1} C(p) \\ &= (n+0.5)\ln n - n + 1 - \left(\sum_{p=1}^{\infty} C(p) - \sum_{p=n}^{\infty} C(p)\right) \\ &= (n+0.5)\log n - n + 1 - D + r_n. \end{aligned}$$

(j) Hence

$$\begin{aligned} n! &= \exp S(n) \\ &= \exp\left((n+0.5)\log n - n + 1 - D + r_n\right) \end{aligned}$$

$$= \quad Kn^{(n+0.5)}\exp(-n+r_n)$$

where $K = \exp(1-D)$ which is a constant.

(k) According to Courant [18, p. 225] Wallis' formula for π is

$$\pi = \lim_{n\to\infty} \frac{2^{4n}(n!)^4}{n(2n!)^2}.$$

There is no time to pause and prove this as we are too close to wrapping up the proof of Stirling's formula. If you want to stop and sniff the roses, you can go back and check this out later.

(l) We have shown by (j) that

$$\lim_{n\to\infty} \frac{n!}{Kn^{n+0.5}\exp(-n)} = 1$$

since, by (h),

$$\lim_{n\to\infty} r_n = 0.$$

Thus, using Wallis' formula, we have

$$\begin{aligned}\pi &= \lim_{n\to\infty}\left(\frac{2^{4n}\left(Kn^{n+0.5}\exp(-n)\right)^4}{n\left(K(2n)^{2n+0.5}\exp(-2n)\right)^2}\right)\\ &= \frac{K^2}{2}.\end{aligned}$$

Thus,

$$K = \sqrt{2\pi}.$$

(m) Hence as $n \to \infty$

$$n! \sim \sqrt{2\pi}n^{n+0.5}\exp(-n)$$

which is Stirling's formula.

(n) We can use Stirling's approximation to approximate $n!$ when n is large. Thus,

$$\begin{aligned}\binom{200}{100} &= \frac{200!}{(100!)^2}\\ &\approx \frac{\sqrt{2\pi}200^{200.5}\exp(-200)}{\left(\sqrt{2\pi}100^{100.5}\exp(-100)\right)^2}\end{aligned}$$

$$= \frac{2^{200}}{10\sqrt{\pi}}$$
$$\approx 9.066 \times 10^{58} \text{ by my calculator.}$$

`Excel` gave me $\binom{200}{100} = 9.05485 \times 10^{58}$.

(2) These results show the important links between probability and classical analysis.

(a) There are different ways of getting at the number e; this one is via limits of sequences. An entertaining history of the number e is presented in Maor [43].

i. We have

$$\begin{aligned} e_n &= \left(1 + \frac{1}{n}\right)^n \\ &= \sum_{k=0}^{n} \binom{n}{k} \left(\frac{1}{n}\right)^k \\ &= \sum_{k=0}^{n} \frac{1}{k!} \frac{n(n-1)(n-2)\dots(n-k+1)}{n^k} \\ &= \sum_{k=0}^{n} \frac{1}{k!} \prod_{j=1}^{k-1} \left(1 - \frac{j}{n}\right). \end{aligned}$$

ii. Thus, since $n < n+1$,

$$\begin{aligned} 0 &< e_n \\ &< \sum_{k=0}^{n} \frac{1}{k!} \prod_{j=1}^{k-1} \left(1 - \frac{j}{n+1}\right) \\ &< \sum_{k=0}^{n+1} \frac{1}{k!} \prod_{j=1}^{k-1} \left(1 - \frac{j}{n+1}\right) \\ &= e_{n+1}. \end{aligned}$$

iii. Also,

$$\begin{aligned} & e_n \\ < & \sum_{k=0}^{n} \frac{1}{k!} \prod_{j=1}^{k-1} \left(1 - \frac{j}{n}\right) \end{aligned}$$

$$\begin{aligned}
&< \sum_{k=0}^{n} \frac{1}{k!} \\
&< 1 + 1 + 2^{-1} + 2^{-2} + \ldots + 2^{-(n-1)} \\
&< 1 + \sum_{k=0}^{\infty} 2^{-k} \\
&= 3.
\end{aligned}$$

iv. Since

$$0 < e_1 < e_2 < \ldots < 3,$$

it follows that $\lim_{n\to\infty} e_n$ exists and < 3. We call this limit e.

(b) The proofs of these results provide good experience in developing your techniques for dealing with inequalities.

i. The method here is much like the one above to find e.

$$\begin{aligned}
&\left(1 - \frac{1}{n^2}\right)^n \\
&= \sum_{k=0}^{n} \binom{n}{k} \frac{(-1)^k}{n^{2k}} \\
&= 1 + \sum_{k=1}^{n} \frac{(-1)^k}{k!\, n^k} \prod_{j=1}^{k-1} \left(1 - \frac{j}{n}\right) \\
&= 1 + \frac{1}{n} \sum_{k=1}^{n} \frac{(-1)^k}{k!\, n^{k-1}} \prod_{j=1}^{k-1} \left(1 - \frac{j}{n}\right) \\
&= 1 + n^{-1} S_n \text{ say.}
\end{aligned}$$

We know that

$$|S_n| < \sum_{k=1}^{\infty} (1/k!) < 3$$

(from above) and hence

$$\lim_{n\to\infty} \left(1 - \frac{1}{n^2}\right)^n = 1.$$

ii. This result now follows easily from those above.

$$\begin{aligned}\lim_{n\to\infty}\left(1-\frac{1}{n}\right)^n &= \lim_{n\to\infty}\frac{\left(1-\frac{1}{n^2}\right)^n}{\left(1+\frac{1}{n}\right)^n}\\ &= e^{-1}.\end{aligned}$$

iii. Suppose that n is a natural number and $n < x \le n+1$. Then

$$\left(1-\frac{1}{n}\right)^{n+1} < \left(1-\frac{1}{x}\right)^x < \left(1-\frac{1}{n+1}\right)^n$$

$$\begin{aligned}&\left(1-\frac{1}{n}\right)^n\left(1-\frac{1}{n}\right)\\ <\ &\left(1-\frac{1}{x}\right)^x\\ <\ &\left(1-\frac{1}{n+1}\right)^{n+1}\left(1-\frac{1}{n+1}\right)^{-1}.\end{aligned}$$

Now let x (and hence n) increase without bound. Then, by the previous result,

$$\lim_{x\to\infty}\left(1-\frac{1}{x}\right)^x = e^{-1}.$$

iv. For $\mu > 0$, let $x = n/\mu$. Then using the previous result,

$$\lim_{n\to\infty}\left(1-\frac{\mu}{n}\right)^n = \lim_{x\to\infty}\left(\left(1-\frac{1}{x}\right)^x\right)^\mu = e^{-\mu}.$$

(c) Since X has a binomial distribution with parameters n and p, we have, for $k = 0, 1, 2, \ldots, n$,

$$P(X = k) = \binom{n}{k} p^k (1-p)^{n-k}$$

and we are considering $np = \mu$.

i. Thus, we have

$$\begin{aligned} & P(X=k) \\ = & \binom{n}{k}\left(\frac{\mu}{n}\right)^k\left(1-\frac{\mu}{n}\right)^{n-k} \\ = & \frac{n(n-1)\ldots(n-k+1)}{n^k\left(1-\frac{\mu}{n}\right)^k}\left(\frac{\mu^k}{k!}\right)\left(1-\frac{\mu}{n}\right)^n. \end{aligned}$$

ii. Now let $n \to \infty$ but keep k fixed; thus,

$$\begin{aligned} & \lim_{n\to\infty} P(X=k) \\ = & \lim_{n\to\infty} \frac{\prod_{j=1}^{k-1}\left(1-\frac{j}{n}\right)}{\left(1-\frac{\mu}{n}\right)^k}\left(\frac{\mu^k}{k!}\right)\left(1-\frac{\mu}{n}\right)^n \\ = & \frac{\exp(-\mu)\mu^k}{k!}. \end{aligned}$$

iii. Now turn this limit result into an approximation for large n. If $n = 100$ and $p = 0.02$ then $\mu = np = 2$ and hence

$$\begin{aligned} P(X \le 2) &\approx \exp(-\mu)\sum_{k=0}^{2}\frac{\mu^k}{k!} \\ &= \exp(-2)\sum_{k=0}^{2}\frac{2^k}{k!} \\ &\approx 0.6766 \text{ by my calculator.} \end{aligned}$$

By comparison, `Excel` gave $P(X \le 2) = 0.676685622$.

(3) The weak law of large numbers (WLLN) is one of the simplest yet general limit theorems. The first version of this result was established by Jacob Bernoulli in 1713. Recall that if $\{X_1, X_2, \ldots\}$ is a sequence of random variables such that $E(X_i)$ and $\mathrm{Var}(X_i)$ exist, and $\{c_1, c_2, \ldots\}$ is a sequence of constants, then

$$E(\sum_{i=1}^{n} c_i X_i) = \sum_{i=1}^{n} c_i E(X_i)$$

$$\text{Var}(\sum_{i=1}^{n} c_i X_i) = \sum_{i=1}^{n} c_i \text{Var}(X_i)\,.$$

(a) $E(\overline{X}) = n^{-1}\sum_{i=1}^{n} E(X_i) = \mu$.
(b) $\text{Var}(\overline{X}) = n^{-1}\sum_{i=1}^{n} \text{Var}(X_i) = \sigma^2$.
(c) Chebyshev's inequality for a r.v. Y is

$$P(|Y - \mu_Y| \geq \epsilon) \leq \frac{\sigma_Y^2}{\epsilon^2}\,.$$

Apply this to $Y = \overline{X}$. Then

$$P(|\overline{X} - \mu_{\overline{X}}| \geq \epsilon) \leq \frac{\sigma_{\overline{X}}^2}{\epsilon^2}\,.$$

$$P(|\overline{X} - \mu| \geq \epsilon) \leq \frac{\sigma^2}{n\epsilon^2}\,.$$

(d) Let $n \to \infty$; then for any $\epsilon > 0$,

$$\lim_{n\to\infty} P(|\overline{X} - \mu| \geq \epsilon) = 0.$$

(e) The significance of this result is that if we have a very large random sample from a population, then it is unlikely that the sample mean ($\overline{X}$) will be very different from the population mean (μ).

(4) In this exercise, we explore the algebra which allows us to manipulate characteristic functions. These results will be useful in subsequent problems.

(a) We have

$$\begin{aligned}\phi_{S_n}(t) &= E\exp(itS_n)\\ &= E\exp(i\sum_{k=1}^{n} X_k t)\\ &= E\prod_{k=1}^{n}\exp(iX_k t)\\ &= \prod_{k=1}^{n} E\exp(iX_k t) \text{ by the independence of the rv}\end{aligned}$$

$$= \prod_{k=1}^{n} \phi_{X_k}(t).$$

(b) We apply the previous result:

$$\begin{aligned} \phi_{S_n}(t) &= \prod_{k=1}^{n} \phi_{X_k}(t) \\ &= \prod_{k=1}^{n} \phi_X(t) \text{ because each rv has the same} \\ &\quad \text{distribution as } X \\ &= (\phi_X(t))^n . \end{aligned}$$

(c) Since $Y = aX + b$, we have

$$\begin{aligned} \phi_Y(t) &= E\exp(itY) \\ &= E\exp(itaX + itb) \\ &= (E\exp(itaX))\,(E\exp(itb)) \\ &= \exp(itb)\phi_X(at). \end{aligned}$$

(5) Let $Z \sim N(0,1)$. To find the c.f. of Z is surprisingly difficult. Here are several proofs of

$$\phi_Z(t) = \exp(-t^2/2).$$

Each proof involves interchanging limiting processes in some way or other. These interchanges can be justified by the fact that, for any continuous r.v. X, we have

$$\begin{aligned} |\phi_X(t)| &= \left|\int_{\Re} \exp(itu) f_X(u)\,du\right| \\ &\le \int_{\Re} |\exp(itu)| f_X(u)\,du \\ &\le \int_{\Re} f_X(u)\,du \\ &= 1. \end{aligned}$$

Therefore, by de la Vallée Poussin's test [71, p. 72], the integral which defines ϕ_X is absolutely convergent and uniformly convergent for $t \in \Re$.

Proof 1: This proof uses countour integration of a function of a complex variable. Thus, it requires an acquaintance with complex analysis. See Stoyanov, Mirazchiiski, Ignatov and Tanushev [65, pp. 129–130].

$$\begin{aligned}
\phi_Z(t) &= \frac{1}{\sqrt{2\pi}} \int_{\Re} \exp(itu) \exp(-u^2/2)\, du \\
&= \frac{1}{\sqrt{2\pi}} \int_{\Re} \exp(itu - u^2/2)\, du \\
&= \frac{\exp(-t^2/2)}{\sqrt{2\pi}} \int_{\Re} \exp(-(u - it)^2/2)\, du \\
&= \frac{\exp(-t^2/2)}{\sqrt{2\pi}} \int_{L(t)} \exp(-z^2/2)\, du
\end{aligned}$$

where $L(t)$ is the line in the complex plane

$$L(t) = \{z = u - it : -\infty < u < \infty\}.$$

The classic approach in complex analysis to calculating this last mentioned integral is by contour integration. The function $\exp(-z^2/2)$ is an entire function and hence the integral of this function around a suitable closed curve is zero. Let $x > 0$ and consider the closed curve C which is the boundary of the rectangle with vertices in the complex plane at

$$-x, x, x - it, -x - it.$$

Then

$$\int_C \exp(-z^2/2)\, dz = 0.$$

Let

$$\begin{aligned}
I_1(x) &= \int_{-x}^{x} \exp(-z^2/2)\, dz \\
I_2(x) &= \int_{x}^{x-it} \exp(-z^2/2)\, dz \\
I_3(x) &= \int_{-x-it}^{-x} \exp(-z^2/2)\, dz \\
I_4(x) &= \int_{-x-it}^{x-it} \exp(-z^2/2)\, dz
\end{aligned}$$

Hence

$$I_4(x) = I_1(x) + I_2(x) + I_3(x).$$

We consider each of the three integrals on the left hand side in turn.
We know that

$$\lim_{x\to\infty} I_1(x) = \sqrt{2\pi}.$$

For $I_2(x)$, let $z = x + i\theta$. Then

$$\begin{aligned} |I_2(x)| &= |(-i)\int_0^t \exp(-x^2/2)\exp(i\theta x)\exp(\theta^2/2)\,d\theta| \\ &\leq \exp(-x^2/2)\int_0^t \exp(\theta^2/2)\,d\theta \end{aligned}$$

and hence

$$\lim_{x\to\infty} I_2(x) = 0;$$

similarly

$$\lim_{x\to\infty} I_3(x) = 0.$$

Thus

$$\lim_{x\to\infty} \frac{1}{\sqrt{2\pi}}\int_{-x-it}^{x-it} \exp(-z^2/2)\,dz = 1$$

and hence

$$\begin{aligned} \phi_Z(t) &= \frac{\exp(-t^2/2)}{\sqrt{2\pi}}\int_{L(t)} \exp(-z^2/2)\,dz \\ &= \exp(-t^2/2). \end{aligned}$$

Proof 2: This proof depends on interchanging the order of the operation of integration over an infinite interval and the operation of infinite summation. See Shiryaev [60, p. 277].

$$\phi_Z(t) = \frac{1}{\sqrt{2\pi}}\int_{\Re} \exp(itu)\exp(-u^2/2)\,du$$

$$
\begin{aligned}
&= \frac{1}{\sqrt{2\pi}} \int_{\Re} \sum_{k=0}^{\infty} \frac{(itu)^k}{k!} \exp(-u^2/2)\, du \\
&= \frac{1}{\sqrt{2\pi}} \sum_{k=0}^{\infty} \frac{(it)^k}{k!} \int_{\Re} u^k \exp(-u^2/2)\, du.
\end{aligned}
$$

Now for $p \in \{0, 1, 2, 3, \ldots\}$ we have [50, p. 25]

$$
\int_{\Re} u^p \exp(-u^2/2)\, du = \begin{cases} \dfrac{(2k)!}{k!\,2^k} & p = 2k \\ 0 & p = 2k+1. \end{cases}
$$

The case $p = 2k$ can be easily proved using induction and integration by parts; the other case follows from the fact that the integrand is an odd function. Thus,

$$
\begin{aligned}
\phi_Z(t) &= \sum_{k=0}^{\infty} \frac{(-t^2)^k}{(2k)!} \int_{\Re} u^{2k} \exp(-u^2/2)\, du \\
&= \sum_{k=0}^{\infty} \frac{(-t^2)^k}{(2k)!} \frac{(2k)!}{2^k k!} \\
&= \sum_{k=0}^{\infty} \frac{(-t^2/2)^k}{k!} \\
&= \exp(-t^2/2)\,.
\end{aligned}
$$

Proof 3: This proof depends on interchanging the order of the operation of differentiation and the operation of integration. See Durrett [22, pp. 66–67].

$$
\begin{aligned}
\phi_Z(t) &= \frac{1}{\sqrt{2\pi}} \int_{\Re} \exp(itu) \exp(-u^2/2)\, du \\
&= \frac{1}{\sqrt{2\pi}} \int_{\Re} \cos(tu) \exp(-u^2/2)\, du \\
&\quad + \frac{i}{\sqrt{2\pi}} \int_{\Re} \sin(tu) \exp(-u^2/2)\, du \\
&= \frac{1}{\sqrt{2\pi}} \int_{\Re} \cos(tu) \exp(-u^2/2)\, du\,;
\end{aligned}
$$

$$
\phi'(t) = \frac{(-1)}{\sqrt{2\pi}} \int_{\Re} u \sin(tu) \exp(-u^2/2)\, du
$$

$$= \quad -t\phi_Z(t) \text{ after integration by parts.}$$

So we must solve the initial value problem

$$\begin{aligned} \phi'_X(t) &= -t\phi_Z(t) \\ \phi_Z(0) &= 1. \end{aligned}$$

This leads to

$$\phi_Z(t) = \exp(-t^2/2).$$

Proof 4: An interesting probabilistic proof is given by Breiman [10, pp. 185–186].

(6) We work through this proof of the CLT following the steps provided.

(a) The function $F_Z(t)$ is defined for all $t \in \Re$ and is continuously differentiable for all $t \in \Re$ because

$$F'_Z(t) = \frac{1}{\sqrt{2\pi}} \exp(-t^2/2);$$

a fortiori, $F_Z(t)$ is continuous for all $t \in \Re$.

(b) We want to prove that

$$(\forall t \in \Re)(\lim_{n\to\infty} F_n(t) = F_Z(t)).$$

Since $F_Z(t)$ is continuous for all $t \in \Re$, we are seeking to prove that $F_n \Rightarrow F_Z$.

(c) We have seen above that the c.f. of Z is

$$\phi_Z(t) = \exp(-t^2/2)$$

which is continuous throughout $\Re$. By Theorem 5.1, it suffices to show that

$$(\forall t \in \Re)(\lim_{n\to\infty} \phi_{Z_n}(t) = \exp(-t^2/2)).$$

(d) If $x = n/\lambda_n$ then

$$\begin{aligned} \left(1 + \frac{\lambda_n}{n}\right)^n &= \left(1 + \frac{1}{x}\right)^{x\lambda_n} \\ &= \left(\left(\left(1 + \frac{1}{x}\right)^x\right)^{\lambda_n/\lambda}\right)^\lambda \end{aligned}$$

and hence,

$$\lim_{n\to\infty}\left(1+\frac{\lambda_n}{n}\right)^n = \exp(\lambda).$$

(e) Note that

$$\begin{aligned} Z(n) &= \frac{S(n)-ES(n)}{\sigma_{S(n)}} \\ &= \frac{\sum_{i=1}^{n}(X_i-\mu)}{\sigma_{S(n)}} \\ &= \frac{\sum_{i=1}^{n} Y_i}{\sigma\sqrt{n}} \text{ (because Var}S(n) = n\text{Var}X = n\sigma^2). \end{aligned}$$

Now use results concerning the algebra of characteristic functions in problem 4 above. Then

$$\begin{aligned} \phi_{Z(n)}(t) &= \phi_{\sum Y_i}\left(\frac{t}{\sigma\sqrt{n}}\right) \\ &= \left(\phi_Y\left(\frac{t}{\sigma\sqrt{n}}\right)\right)^n. \end{aligned}$$

(f) Since $\text{Var}\,(Y) = \text{Var}\,(X-\mu) = \text{Var}\,(X) = \sigma^2$,

$$\begin{aligned} \phi_Y\left(\frac{t}{\sigma\sqrt{n}}\right) &= 1-\sigma^2\left(\frac{t^2}{2\sigma^2 n}\right)+K\left(\frac{t}{\sigma\sqrt{n}}\right) \\ &= 1-\left(\frac{t^2}{2n}\right)+K\left(\frac{t}{\sigma\sqrt{n}}\right) \\ &= 1+\frac{1}{n}\left(-\frac{t^2}{2}+nK\left(\frac{t}{\sigma\sqrt{n}}\right)\right). \end{aligned}$$

Now

$$\lim_{t\to 0}\frac{K(t)}{t^2} = 0.$$

So $K(t) = \epsilon(t)t^2$ where $\lim_{t\to 0}\epsilon(t) = 0$. Hence

$$\begin{aligned} nK\left(\frac{t}{\sigma\sqrt{n}}\right) &= n\epsilon\left(\frac{t}{\sigma\sqrt{n}}\right)\left(\frac{t^2}{\sigma^2 n}\right) \\ &= \epsilon\left(\frac{t}{\sigma\sqrt{n}}\right)\left(\frac{t^2}{\sigma^2}\right) \\ &= H(t,n) \end{aligned}$$

where $\lim_{n\to\infty} H(t,n) = 0$. Therefore,

$$\phi_Y\left(\frac{t}{\sigma\sqrt{n}}\right) = 1 + \frac{1}{n}\left(\frac{-t^2}{2} + H(t,n)\right)$$

where $\lim_{n\to\infty} H(t,n) = 0$.

(g) Using 6e, 6f, 6d respectively, we obtain

$$\begin{aligned}\lim_{n\to\infty} \phi_{Z(n)}(t) &= \lim_{n\to\infty}\left(\phi_Y\left(\frac{t}{\sigma\sqrt{n}}\right)\right)^n \\ &= \lim_{n\to\infty}\left(1 + \frac{1}{n}\left(-\frac{t^2}{2} + H(t,n)\right)\right)^n \\ &= \exp(-t^2/2).\end{aligned}$$

According to 6c, we have proved the CLT.

(7) For $i = 1, 2, \ldots, n$, the distribution of X_i is the same as the distribution of X where

$$P(X = t) = 1/6 \quad (t = 1, 2, 3, 4, 5, 6).$$

(a) $\mu = E(X) = \sum_{t=1}^{6} t/6 = 7/2\,(i = 1, 2, 3, \ldots, n)$.
(b) $E(X^2) = \sum_{t=1}^{6} t^2/6 = 91/6\,(i = 1, 2, 3, \ldots, n)$.
$\sigma^2 = (91/6) - (7/2)^2 = 35/12$.
(c) $E(\overline{X}) = E(X) = 7/2$ and $\text{Var}(\overline{X}) = \text{Var}(X)/n = 35/12n$.
(d) $n = 25$; $E(\overline{X}) = 3.5$; $\text{Var}(\overline{X}) = (0.3416)^2$.

i. According to Chebyshev's theorem,

$$P(|\overline{X} - 3.5| \geq 0.5) \leq \frac{\sigma^2}{n(0.5)^2} = 0.4667.$$

ii. According to the CLT,

$$\overline{X} \sim N(\mu, \sigma^2/n) \text{ approximately;}$$

$$Z = \frac{\overline{X} - \mu}{\sigma/\sqrt{n}} = \frac{\overline{X} - 3.5}{0.3416} \sim N(0,1) \text{ approximately.}$$

So,

$$P(|\overline{X} - 3.5| \geq 0.5) = P(|Z| \geq 1.46) \approx 0.144.$$

iii. The first estimate from Chebyshev's theorem is an upper estimate of the required probability whereas the second estimate from the CLT is an approximation. We do not know whether the required probability exceeds 0.144 or does not; nor do we know how far it is from 0.144.
It is interesting to note that Chebyshev's inequality may yield results which, although correct, are not very informative. For example, if we wanted to estimate

$$P(|\overline{X} - 3.5| \geq 0.1)$$

in this problem, using Chebyshev's inequality would lead to

$$P(|\overline{X} - 3.5| \geq 0.1) \leq \frac{\sigma^2}{n(0.1)^2} = 11.67$$

which is not a very useful piece of information since all probabilities do not exceed 1.

(e) Now repeat the above calculations for $n = 500$.
According to Chebyshev's theorem,

$$P(|\overline{X} - 3.5| \geq 0.5) \leq \frac{\sigma^2}{n(0.5)^2} = 0.0233.$$

Applying the CLT leads to

$$P(|\overline{X} - 3.5| \geq 0.5) = P(|Z| \geq 6.55) \approx 0.00.$$

(8) For $i = 1, 2, 3, \ldots, 52$, let X_i denote the income in week i. Then $A = \sum_{i=1}^{52} X_i$ and we have

$$E(A) = (52)(2000) = 104,000$$

and

$$\sigma_A = 250\sqrt{52} = 1802.78.$$

According to the CLT, if we make enough assumptions,

$$Z = \frac{A - 104,000}{1802.78} \sim N(0,1) \text{ approximately.}$$

(a) Hence $P(A \geq 110,000) \approx P(Z \geq 3.32) = 0.0004$.

(b) From tables of the standard Normal distribution,

$$\begin{aligned} 0.90 &= P(-1.645 \leq Z \leq 1.645) \\ &= P(A \in [104000 \pm (1.645)(1802.78)]) \\ &= P(A \in [101034.42, 106965.57]). \end{aligned}$$

Thus it seems reasonably likely that if next year is going to be like the last year then the annual income is likely to fall in the range \$ 101,034.42 – \$ 106,965.57.

(c) There are some obvious difficulties associated with this application of the CLT.

First, if income in the small business experiences seasonal influences, then the assumption that the variables X_i are identically distributed would not be valid.

Second, as the data is time series data (i.e. a sequence of random variables indexed by time), it is often the case that the r.v. are not independent of each other. High sales in one week may have flow on effects to the following week.

On the other hand, the version of the CLT which you have proved earlier in this chapter is the classical version. There are many other versions in which pure mathematicians have proved the CLT with weaker assumptions in order to deal with the practical difficulties suggested by applications such as this one. This is a nice illustration of practical problems posing challenges for the pure mathematician.

(9) Suppose that we choose a number X at random from the set

$$\{1, 2, 3, \ldots, 44, 45\}.$$

Then

$$\begin{aligned} E(X) &= \sum_{t=1}^{45} (t/45) = 23 \\ E(X^2) &= \sum_{t=1}^{45} (t^2/45) = 4186 \\ \mathrm{Var}\,(X) &= E(X^2) - E(X)^2 = (60.47)^2. \end{aligned}$$

Now using the CLT approximation, we get

$$X_1 + X_2 + X_3 + X_4 + X_5 + X_6 \sim N(138, 6(60.47)^2) \text{ approximately}$$

and hence

$$Z = \frac{(X_1 + \ldots + X_6) - 138}{60.47\sqrt{6}} \sim N(0,1) \text{ approximately.}$$

Now $P(Z \geq 0.8416) \approx 0.2$; so,

$$P(X_1 + X_2 + \ldots + X_6 \geq 262) \approx 0.20.$$

This calculation can be described as "quick and dirty". However it could be done on the back of an envelope to give you some idea of the number T. You could check the calculation by simulating the selection of six numbers from $\{1, 2, \ldots, 45\}$ and then estimating the probability that the sum exceeds 262 from the simulation.

(10) Assume that we have n components in stock with lifetimes

$$\{X_1, X_2, \ldots, X_n\}.$$

Then we regard

$$\{X_1, X_2, \ldots, X_n\}$$

as i.i.d random variables.

(a) The recommendation is that

$$P(X_1 + X_2 + \ldots + X_n \geq 1000) \geq 0.99$$

$$P\left(\frac{(X_1 + X_2 + \ldots + X_n) - 50n}{15\sqrt{n}} \geq \frac{1000 - 50n}{15\sqrt{n}}\right) \geq 0.99.$$

Hence, using the table of the standard Normal distribution,

$$\frac{1000 - 50n}{15\sqrt{n}} \leq -2.3263$$

$$50n - (2.3263)15\sqrt{n} - 1000 \geq 0$$

and this is a quadratic equation in $\sqrt{n}$. So

$$\begin{aligned} \sqrt{n} &\geq \frac{6.9789 + \sqrt{(6.9789)^2 + 8000}}{20} \\ n &\geq 24. \end{aligned}$$

(b) If p were reduced from 0.99 to 0.95 then, in the above calculation we replace 2.3263 by 1.6449 and obtain $n \geq 23$. It is interesting to note that such a major increase in the risk leads to little change in stock requirements. Unless the price of a component is extrordinarily high, there would appear to be little benefit in increasing the risk of system failure. Here we see the power of mathematical modelling in being able to answer "what if ..." questions.

(c) A mathematician who has solved the previous two parts of this problem could not resist looking for some general formula to automate decision making.

$$P(X_1 + \ldots + X_n \geq H) \geq p$$

$$P\left(\frac{(X_1 + \ldots + X_n) - n\mu_X}{\sigma_X\sqrt{n}} \geq \frac{H - n\mu_X}{\sigma_X\sqrt{n}}\right) \geq p$$

$$\frac{H - n\mu_X}{\sigma_X\sqrt{n}} \leq -z_p \quad \text{where } P(Z \leq z_p) = p$$

$$\mu_X n - z_p\sigma_X\sqrt{n} - H \geq 0$$

$$n \geq \left(\frac{z_p\sigma_X + \sqrt{z_p^2\sigma_X^2 + 4\mu_X H}}{2\mu_X}\right)^2.$$

Again we see the power of mathematical modelling. Armed with this general formula, the quality engineer could make recommendations for many different situations.

(11) Let $Z \sim N(0,1)$. Then

$$\begin{aligned} P(Z > 0.4307) &= 1/3 \quad \text{(by standard Normal tables)} \\ &= \lim_{n\to\infty} P\left(\frac{S(n)}{\sqrt{n}} > 1\right) \end{aligned}$$

$$= \lim_{n\to\infty} P(\frac{S(n)}{\sigma\sqrt{n}} > \frac{1}{\sigma})$$
$$= P(Z > \frac{1}{\sigma}) \quad \text{(by CLT)}.$$

So $\sigma = 1/0.4316$; $\sigma^2 = 5.368$.

(12) The method of characteristic functions was outlined in our proof of the central limit theorem. Essentially to prove weak convergence of a sequence of distribution functions, we prove convergence of the associated characteristic functions.

(a) For all $t \in \Re$, $F_Z(t)$ is continuous. So, we are being asked to prove that, as $n \to \infty$, $F_{Z(n)}$ converges weakly to F_Z.
Let $\phi_{Z(n)}$ be the c.f. of $Z(n)$. By the method of c.f., it suffices to prove that

$$\lim_{n\to\infty} \phi_{Z(n)}(t) = \exp(-t^2/2) \quad (t \in \Re).$$

$$\begin{aligned}
\phi_{X(n)}(t) &= E(\exp(iX(n)t) \\
&= \sum_{k=0}^{\infty} \exp(ikt)\exp(-n)n^k/k! \\
&= \exp(-n)\sum_{k=0}^{\infty} \exp(ikt)n^k/k! \\
&= \exp(-n + n\exp(it)).
\end{aligned}$$

$$\begin{aligned}
\phi_{Z(n)}(t) &= \exp(-it\sqrt{n})\phi_{X(n)}(t/\sqrt{n}) \\
&= \exp(-it\sqrt{n})\exp(-n + n\exp(it/\sqrt{n})) \\
&= \exp(-it\sqrt{n} - n + n\exp(it/\sqrt{n})) \\
&= \exp(-t^2/2 + \epsilon_n) \text{ where } \lim_{n\to\infty} \epsilon_n = 0\,.
\end{aligned}$$

Thus,

$$\lim_{n\to\infty} \phi_{Z(n)}(t) = \exp(-t^2/2) \quad (t \in \Re).$$

and hence, as $n \to \infty$, $F_{Z(n)}$ converges weakly to F_Z. Thus,

$$(\forall t \in \Re)(\lim_{n\to\infty} F_{Z(n)}(t) = F_Z(t)).$$

(b) Suppose that $X \sim \text{Poisson}(n = 20)$. Then, using the approximation suggested by the first part of this exercise,

$$Z = \frac{X - 20}{\sqrt{20}} \sim N(0, 1) \text{ approximately.}$$

More correctly,

$$\begin{aligned} P(15 \le X \le 20) &= P(14.5 \le X \le 20.5) \\ &\approx P(-1.23 \le Z \le 0.11) \\ &= 0.4345. \end{aligned}$$

Notice that, because we are approximating the distribution of a discrete r.v. X by the distribution of a continuous r.v. Z, we are approximating the area under part of a histogram by the area under part of a continuous density function. Hence we use $P(X \in [14.5, 20.5])$ rather than $P(X \in [15, 20])$.

Chapter 6

Random walks

You will notice that the style of arguments used here are very different from the styles of arguments which you encounter in calculus and algebra. In those subjects there is a mechanical approach to many problems

(... differentiate both sides, solve for dy/dx, put $dy/dx = 0$ and solve ...).

On the other hand, in this chapter we set out our arguments in a more descriptive fashion; these sorts of arguments bring your analytical skills up to a higher plane. Also, you see that mathematics deals with ideas, concepts and arguments — not merely calculation.

(1) This question explores the definition of a path.

(a) There are many paths from $(0,0)$ to $(13,7)$. One is the polygonal path which joins the following points:

$$(0,0),(1,1),(2,2),(3,3),(4,4),(5,5),(6,6),$$

$$(7,7),(8,8),(9,9),(10,10),(11,9),(12,8),(13,7).$$

Try as you may, you cannot find a path from $(0,0)$ to $(13,8)$. The reason is found in the next part of this question.

(b) Consider a path from $(0,0)$ to (n,x). Suppose that p is the number of steps in which $\epsilon_k = +1$ and q is the number of steps in which $\epsilon_k = -1$. Then

$$n = p + q \text{ and } x = p - q.$$

Conversely, if there exist non-negative integers p, q such that

$$n = p + q \text{ and } x = p - q$$

then we can construct a path consisting of p consecutive upward steps and q consecutive downward steps which will start at $(0, 0)$ and end at (n, x).

In particular, we now see why you cannot find a path from $(0, 0)$ to $(13, 8)$: there do not exist non-negative integers p, q such that

$$13 = p + q \text{ and } 8 = p - q\,.$$

(2) As we will see, there is an art to counting paths.

(a) The path has n steps $\epsilon_k (k = 1, 2, \ldots, n)$ and, for each of the n steps there are two possibilities, $+1$ or -1. Thus there are 2^n possible paths.

(b) We saw above that there is a path from $(0, 0)$ to (n, x) if and only if there exist non-negative integers p, q such that

$$n = p + q \text{ and } x = p - q\,.$$

Of the n steps, choose the p for which $\epsilon_k = +1$ and, for the remaining steps, put $\epsilon_k = -1$. This describes the totality of such paths; thus the number of paths is $\binom{p+q}{p}$. If it is not the case that $n = p + q$ and $x = p - q$ then there are no admissable paths. Thus, the number of paths from $(0, 0)$ to (n, x) is given by

$$N_{n,x} = \begin{cases} \dbinom{p+q}{p} & \text{if } n = p + q \text{ and } x = p - q \\ \\ 0 & \text{otherwise.} \end{cases}$$

(c) The number of paths from (a, b) to (n, x) is the number of paths from $(0, 0)$ to $(n-a, x-b)$ and hence is equal to $N_{n-a,x-b}$. We see this by shifting the origin from $(0, 0)$ to (a, b).

(3) Let

$$A = (a, \alpha) = (n_1, x_1), (n_1 + 1, x_2), \ldots, (n_1 + k, x_{k+1}) = (b, \beta) = B$$

be a path from A to B which touches or crosses the x-axis; i.e. is an element of S_1. Define j to be the smallest integer ≥ 1 such that $x_j = 0$.
We associate this path in S_1 with the following path in S_2:

$$A' = (a, -\alpha) = (n_1, -x_1), (n_1 + 1, -x_2), \ldots, (n_1 + j - 1, 0),$$

$$(n_1 + j, x_{j+1}) \ldots (n_1 + k, x_{k+1}) = (b, \beta) = B\,.$$

This defines a mapping $\phi : S_1 \to S_2$. A little thought will convince you that this mapping is a bijection. This is the 1–1 correspondence between the elements of S_1 and S_2. Hence $\#S_1 = \#S_2$ as both sets are finite.
A notable aspect of this argument is that we have proved that the number of elements in the set S_1 is equal to the number of elements in the set S_2 without having calculated either number (— yet!). Clever counting arguments such as this is the hallmark of combinatorial mathematics which is now so important in computer science. Feller [25, Chapter 2] gives a classical introduction to this field.

(4) These exercises are routine but designed to reinforce the principles established above.

(a) There is a path from (1,1) to (15,5) if there is a path from $(0, 0)$ to $(14, 4)$. We verify this latter statement by solving

$$14 = p + q \text{ and } 4 = p - q$$

and obtaining $(p, q) = (9, 5)$. Thus there is a path from (1,1) to (15,5); indeed such a path will consist of 14 steps, of which 9 will have $\epsilon_k = +1$ and the remaining 5 will have $\epsilon_k = -1$.

(b) The number of paths from (1,1) to (15,5) will be

$$N_{14,4} = \binom{14}{9} = 2002.$$

(c) The number of paths from (1,1) to (15,5) which touch or cross the horizontal axis equals the number of paths from (1,-1) to (15,5) by the argument used in the reflection principle. This in turn equals the number of paths from (0,0) to (14,6) by translating the origin to (1,-1). Now by solving

$$14 = p + q \text{ and } 6 = p - q$$

and obtaining $(p,q) = (10,4)$. Thus the number of required paths is

$$N_{14,6} = \binom{14}{4} = 1001.$$

(d) The number of paths from (1,1) to (15,5) which do not touch or cross the horizontal axis is therefore, 2002-1001 = 1001.

(5) A path which starts at (0,0) and ends at (n,x) but does not re-visit the x-axis must go through $(1,1)$. So we are interested in the number of paths which start at $(1,1)$, end at (n,x) but do not re-visit the x-axis. We appeal again to the reflection principle. The number of paths which start at $(1,1)$, end at (n,x) but **do** re-visit the x-axis is equal to the number of paths which start at $(1,-1)$ and end at (n,x) and this, in turn, is equal to the number of paths which start at $(0,0)$ and end at $(n-1,x+1)$; i.e. $N_{n-1,x+1}$.
Now the total number of paths from $(1,1)$ to (n,x) is $N_{n-1,x-1}$.
Hence number of paths which start at $(1,1)$, end at (n,x) but do **not** re-visit the x-axis is equal to

$$N_{n-1,x-1} - N_{n-1,x+1}\,.$$

If $n = p + q$ and $x = p - q$ then

$$\begin{aligned} N_{n-1,x-1} - N_{n-1,x+1} &= \binom{p+q-1}{p-1} - \binom{p+q-1}{p} \\ &= \frac{(p+q-1)!}{(p-1)!q!} - \frac{(p+q-1)!}{p!(q-1)!} \\ &= \frac{(p+q-1)!(p-q)}{p!q!} \\ &= \frac{p-q}{p+q}\binom{p+q}{p} \\ &= \frac{x}{n}N_{n,x}\,. \end{aligned}$$

If it is not the case that $n = p+q$ and $x = p-q$ then $N_{n,x} = 0$ and the problem is trivial.

(6) S_n represents the lead of P over Q after the n-th vote is counted.

(a) Note that S_n satisfies the properties of a simple random walk in Definition 6.2.

(b) The event $S_n > 0$ signifies that, after the n-th vote is counted, P is ahead of Q.

(c) To say that throughout the counting of votes P is leading Q is equivalent to saying that the random walk does not re-visit the x-axis. Thus, using the preceeding problem (with $x = 500$ and $n = 2500$) the required probability is

$$\frac{(x/n)N_{n,x}}{N_{n,x}} = \frac{x}{n} = 0.2.$$

(d) In the more general setting $x = p - q$, $n = p + q$ and the required probability is

$$\frac{x}{n} = \frac{p-q}{p+q}.$$

(7) Suppose that $(n+r)/2 \in \{0, 1, \ldots, n\}$. There are 2^n possible paths of n steps and each path is as likely as any other path. Now the number of paths from $(0,0)$ to (n,r) is

$$N_{n,r} = \binom{n}{(n+r)/2}.$$

Thus

$$P(S_n = r) = \binom{n}{(n+r)/2} 2^{-n}.$$

For a non-symmetric, simple random walk, suppose that

$$P(\epsilon_k = +1) = 1 - P(\epsilon_k = -1) = \theta\,.$$

A path from $(0,0)$ to (n,r) will exist if and only if $n = p+q$ and $x = p - q$ (p, q non-negative integers), and, in this case, $p = (n+r)/2$, $q = (n-r)/2$. Such a path will have p steps in which $\epsilon_k = +1$ and q steps in which $\epsilon_k = -1$. The probability of such a path is

$$\binom{n}{(n+r)/2} \theta^{(n+r)/2}(1-\theta)^{(n-r)/2}\,.$$

(8) Limit theorems abound in the study of random walks. (See [55].)

(a) For a symmetric random walk,

$$\begin{aligned} P(S_{2\nu} = 0) &= \binom{2\nu}{\nu} 2^{-2\nu} \\ &= \frac{(2\nu)!}{(\nu!)^2} 2^{-2\nu} \\ &\sim \frac{\sqrt{2\pi}(2\nu)^{2\nu+0.5}\exp(-2\nu)}{(2\pi)\nu^{2\nu+1}\exp(-2\nu)} 2^{-2\nu} \\ &= (\sqrt{\pi\nu})^{-1}. \end{aligned}$$

(b) Using `Excel`, I obtained the following.

ν	$P(S_{2\nu} = 0)$	$(\sqrt{\pi\nu})^{-1}$	Relative error (%)
10	0.176197052	0.178412412	1.26
15	0.144464448	0.145673124	0.84
20	0.125370688	0.126156626	0.63
25	0.112275173	0.112837917	0.50
30	0.102578173	0.103006454	0.42
35	0.095025474	0.095365445	0.36
40	0.088927879	0.089206206	0.31
45	0.083871123	0.084104417	0.28
50	0.079589237	0.079788456	0.25
55	0.075902608	0.076075308	0.23
60	0.072684979	0.072836562	0.21

(9) Let X_i denote the result of the i-th toss.

(a) P(No heads in n throws) $= 0.5^n$. Hence

$$\lim_{n\to\infty} \text{P(No heads in } n \text{ throws)} = 0.$$

Thus, if n is very large, it is extremely unlikely that a head will not have appeared yet. In this sense, a head is bound to turn up eventually.

(b) Suppose that we toss the coin $3n$ times and group the results as

$$(X_1, X_2, X_3)(X_4, X_5, X_6) \ldots (X_{3n-2}, X_{3n-1}, X_{3n}).$$

Now $P(X_{3k-2}, X_{3k-1}, X_{3k}$ is not $HHH) = 7/8 = 0.875$ and hence

$$\lim_{n\to\infty} \text{P(None of the } n \text{ triples is } HHH) = \lim_{n\to\infty} 0.875^n = 0.$$

In this sense, a sequence of 3 consecutive heads is bound to turn up eventually.

(c) The argument in the previous section can be adapted to any given finite sequence of results. Thus, any given finite sequence of heads and tails is bound to turn up eventually.

(d) In a symmetric random walk taking place on the integers, starting at 5 is bound to reach 0 or 10 eventually because there will be a sequence of 11 consecutive steps to the right (or left).

(e) This game is a simple random walk taking place on the integers starting at 0. As in the previous part, it is bound to reach +5 or -5 eventually and the game will come to an end.
This is the study of random walks with barriers, which is an interesting topic with a rich literature.

(10) This problem is the beginning of time series analysis.

(a) For $t \geq 1$,

$$\begin{aligned} Z_t &= Z_{t-1} + e_t \\ &= Z_{t-2} + e_{t-1} + e_t \\ &\vdots \\ &= e_1 + e_2 + \ldots + e_t. \end{aligned}$$

(b) Hence $E(Z_t) = E(e_1 + e_2 + \ldots + e_t) = 0$.
(c) Also, Var $(Z_t) = \text{Var} \sum e_j = t\sigma_e^2$.
(d) If $1 \leq t \leq s$ we have

$$\begin{aligned} \text{Cov}\,(Z_t, Z_s) &= E(Z_t Z_s) - E(Z_t)E(Z_s) \\ &= E(Z_t Z_s) \\ &= E((e_1 + e_2 + \ldots + e_t)(e_1 + e_2 + \ldots + e_s)) \\ &= t\sigma_e^2. \end{aligned}$$

(e) If $1 \leq t \leq s$ we have

$$\begin{aligned} \text{Corr } (Z_t, Z_s) &= \frac{\text{Cov } (Z_t, Z_s)}{\text{SD } (Z_t)\text{SD } (Z_s)} \\ &= \sqrt{t/s}\,. \end{aligned}$$

(f) Thus, $\text{Corr } (Z_t, Z_{t+1}) = \sqrt{t/(t+1)}$ and hence

$$\lim_{t\to\infty} \text{Corr } (Z_t, Z_{t+1}) = 1.$$

If you would like to learn more about time series analysis, see Chatfield [12].

Bibliography

1. Adams, W., *The Life and Times of the Central Limit Theorem*, New York: Kaedmon Publishing Co., 1974.
2. *Kolmogorov in Perspective*, Providence: American Mathematical Society, 2000.
3. Andrews, G., Askey, R. and Roy, R., *Special Functions*, Cambridge: Cambridge University Press, 1998.
4. Apostol, T., *Mathematical Analysis*, Reading, Mass.: Addison-Wesley, 1957.
5. Artin, E., *The Gamma Function,* New York: Holt, Rinehart and Winston, 1964.
6. Australia. Office of the Australian Actuary, *Australian Life Tables, 1985–87*, Canberra: Australian Government Publishing Service, 1991.
7. Bennett, D.J. *Randomness*, Cambridge, Mass: Harvard University Press.
8. Bharucha-Reid, A.T. and Sambandham, M., *Random Polynomials.* Orlando: Academic Press, 1986.
9. Box, G.E.P. and Muller, M.E., A note on the generation of random normal deviates. *Annales of Mathematical Statistics*, v. 29 (1958), pp. 610–611.
10. Breiman, L., *Probability*, Philadelphia: Society for Industrial and Applied Mathematics, 1992.
11. Butzer, P. and Jongmans, F., P. L. Chebyshev (1821–1894). A guide to his life and work. *J. Approx. Theory*, v.96, no. 1 (1999), pp. 111–138.
12. Chatfield, C. *The Analysis of Time Series: An Introduction*, 5th ed. London: Chapman and Hall, 1996.
13. (no author) Pafnutiĭ Lvovich Chebyshev (on the 175th anniversary of his birth), *Theory Probab. Appl.*, v. 41, no. 3 (1996), pp. 519–531.
14. Chebyshev, P.L. et al., *Oeuvres de P.L. Tchebychef*, vols. 1 and 2, New York: Chelsea, 1962.
15. Clark, R., Tattslotto numbers and the perception of randomness. Research Report 251, Department of Mathematics, Monash University, 1995.
16. Clark, R., Are your Tattslotto numbers overdue? Research Report 255, Department of Mathematics, Monash University, 1996.

17.Copi, I.M., *Symbolic Logic*, 5th ed., New York: Macmillan, 1979.
18.Courant, R., *Differential and Integral Calculus*, Vol. I, 2nd ed., London: Blackie & Sons, 1937.
19.Cox, D. and Miller, H., *The Theory of Stochastic Processes*, London: Chapman and Hall.
20.Cramér, H., *Mathematical Methods of Statistics*, Princeton: Princeton University Press, 1946.
21.Doob, J.L., *Measure Theory*, New York: Springer-Verlag, 1993.
22.Durrett, R., *Probability Theory and Examples*, Pacific Grove: Wadsworth, 1991.
23.Epstein, R.A., *The Theory of Gambling and Statistical Logic*, New York: Academic Press, 1977.
24.Erdős, P. and Révész, P., Problems and results on random walks. In *Mathematical Statistics and Probability Theory, Vol. B* (Bad Tatzmannsdorf, 1986) ed. P. Bauer et al., Dordrecht-Boston, MA-London: Reidel, 1987, pp. 59–65.
25.Feller, W., *Introduction to Probability Theory and Its Applications*, vol. 1, 3rd ed. New York: John Wiley & Sons, 1968.
26.Freund, J.E. and Perles, B.M., *Statistics: A First Course,* 7th. ed. Upper Saddle River: Prentice Hall, 1999.
27.Grimmett, G.R. and Stirzaker, D.R., *Probability and Random Processes: Problems and Solutions*, Oxford: OUP, 1991.
28.Hald, A., *Statistical Theory of Sampling Inspection by Attributes*, London: Academic Press, 1981.
29.Halmos, P., *Measure Theory,* Princeton: D. van Nostrand, 1950.
30.Hammersley, J.M. and Handscomb, D.C., *Monte Carlo Methods*, London: Methuen, 1964.
31.Harward, P.M. and Mills, T.M. (1980) How to invert a matrix and probably get the right answer. *Menemui Mathematik*, vol. 2, no.1 pp. 30–36.
32.Henze, N. and Riedwyl, H., *How To Win More: Strategies for Increasing a Lottery Win*, Natick, Mass: A.K. Peters, 1998.
33.Heyde, C.C., Patrick Alfred Pierce Moran 1917-1988. Available URL: `http://www.science.org.au/academy/memoirs/moran.htm`
34.Heyde, C.C. and Seneta, E., *Statisticians of the Centuries.* New York: Springer-Verlag, 2001.
35.Jacod, J. and Protter, P., *Probability Essentials.* Berlin: Springer-Verlag, 2000.
36.Jones, B., *Barry Jones' Dictionary of Biography.* Melbourne: Information Australia, 1996.
37.Kahaner, D., Moler, C., and Nash, S., *Numerical Methods and Software*, Englewood Cliffs: Prentice Hall, 1989.
38.Kendall, M.G. and Moran, P.A.P., *Geometric Probability*, London: Griffin, 1963.
39.Keyfitz, N. and Beekman, J.A. *Demography Through Problems*, New York: Springer-Verlag, 1984.

40.Kolmogorov, A.N., *Foundations of the Theory of Probability*, 2nd English Edition. New York: Chelsea, 1956.

41.Krige, D.G., Statistical applications in mine valuation. *J. Institute Mine Surveyors of South Africa*, vol. 12, no. 2/3 (1962), pp. 1–82.

42.Lukacs, E., *Characteristic Functions*, Second Edition, London: Griffin, 1970.

43.Maor, E., *e: The Story of a Number*, Princeton: Princeton University Press, 1994.

44.Mitra, A., *Quality Control and Improvement,* Second Edition, Upper Saddle River: Prentice-Hall, 1998.

45.Moran, P., *An Introduction to Probability Theory*, Oxford: OUP, 1968.

46.Moran, P.A.P., Calculation of the normal distribution function. *Biometrika*, vol. 67, no. 3 (1980), pp. 675–676.

47.Mosteller, F., *Fifty Challenging Problems in Probability with Solutions*, Reading, Mass: Addison-Wesley, 1965.

48.O'Connor, J.J. and Robertson, E.F., *MacTutor History of Mathematics Archive* [online], 1998.
Available: URL `http://www-history.mcs.st-and.ac.uk/history/`

49.Namboodiri, K. and Suchindran, C., *Life Table Techniques and Their Applications*, Orlando: Academic Press, 1987.

50.Patel, J.K. and Read, C.B., *Handbook of the Normal Distribution*, New York: Marcel Dekker, 1996.

51.Peterson, I., *The Jungles of Randomness: Mathematics and the Edge of Certainty*, London: Penguin Books, 1998.

52.Praetz, P.D., Australian share prices and the random walk hypothesis, *Australian Journal of Statistics*, v. 11, no.3 (1969), pp. 123–139.

53.Råde, L. and Westergren, B., *Beta β Mathematics Handbook*, 2nd ed., Lund: Studentlitterature, 1990.

54.Rényi, A., *Probability Theory,* Budapest: Akadémiai Kiadó, 1970.

55.Révész, P., *Random Walk in Random and Non-Random Environments*, Singapore: World Scientific Press, 1990.

56.Rosenthal, T.S., *A First Look at Rigorous Probability Theory*, Singapore: World Scientific Press.

57.Ross, S., *A First Course in Probability,* 5th ed., Upper Saddle River: Prentice Hall, 1999.

58.Savage, R., Probability inequalities of the Tchebycheff type. *Journal of Research of the National Bureau of Standards – B. Mathematics and Mathematical Physics*, vol. 65B, no.3, July–September (1961), pp. 211–222.

59.Sevastanyov, B., Christyakov, V., and Zubkov, A., *Problems in the Theory of Probability*, Moscow: Mir, 1985.

60.Shiryaev, A.N., *Probability*, 2nd ed., New York: Springer, 1996.

61.Shiryayev, A.N. (ed.) *Selected Works of A.N. Kolmogorov, Vol. II: Probability Theory and Mathematical Statistics*, Dordrecht: Kluwer Academic Publishers, 1992.

62.Spitzer, F., *Principles of Random Walk*, 2nd ed., New York: Springer, 1976.

63. Stewart, J., *Calculus*, 4th ed. Pacific Grove: Brooks/Cole, 1999.
64. Stoyanov, J., *Counterexamples in Probability*, 2nd ed. New York: John Wiley and Sons, 1997.
65. Stoyanov, J., Mirazchiiski, I., Ignatov, Z. and Tanushev, M., *Exercise Manual in Probability Theory*, Dordrecht: Kluwer, 1989.
66. Stuart, A. and Ord, K., *Kendall's Advanced Theory of Statistics*, 6th ed. Vol. 1, London: Edward Arnold, 1994.
67. *The Tattersall's Times*, Issue 303, 26 September 1998.
68. Tuckwell, H.C., *Elementary Applications of Probability Theory*, London: Chapman and Hall, 1988.
69. Tversky, A. and Kahneman, D., Evidential impact of base rates. In *Judgement Under Uncertainty: Heuristics and Biases.* eds. D. Kahneman *et al.* Cambridge: Cambridge University Press, 1982, pp.153–160.
70. Uspensky, J.V., *Introduction to Mathematical Probability*, New York: McGraw-Hill, 1937.
71. Whittaker, E.T. and Watson, G.N., *A Course in Modern Analysis*, 4th ed. Cambridge: Cambridge University Press, 1927.
72. Wimmer, G. W. and Altman, G., *Thesaurus of Univariate Discrete Probability Distributions*, Essen: STAMM, 1999.

Index